AHERA ASBESTOS CONTRACTOR/SUPERVISOR MANUAL

By:

Safety, Health & Ergonomics Branch Electro-Optics,
Environmental & Materials Laboratory
US Environmental Protection Agency
Cooperative Agreement No. CX82760-010-0
Karen Hoffman, EPA Project Officer
Office of Pollution Prevention & Toxic Washington, DC

Published by ehsMaterials.com

January 2015

Table of Contents

The U.S. **Environmental Protection Agency** (EPA) and U.S. **Occupational Safety and Health Administration** (OSHA) both require asbestos-specific training for work involving asbestos .

The OSHA rules apply to all construction work and use a very broad definition of construction.

The EPA rule is based on the Asbestos Hazard Emergency Response Act (AHERA) - Title II of the Toxics Substances Control Act. It establishes a Model Accreditation Program for states. In addition to schools the MAP applies to public and commercial buildings. This term includes residential apartment buildings with more than 10 units in the building.

EPA also has an Asbestos National Emission Standard for Hazardous Air Pollutants (NESHAP) that imposes additional requirements. The NESHAP applies to buildings including residential apartment buildings with more than four units. Some states - apply the MAP training requirements to all buildings covered by the NESHAP.

Information on requirements for specific AHERA training courses

Asbestos Contractor / Supervisor – Initial

EPA Model Accreditation Program: A person must be accredited as a contractor/supervisor to supervise any of the following activities with respect to friable ACBM in a school or public and commercial building: (1) A response action other than a SSSD activity, (2) a maintenance activity that disturbs friable ACBM other than a SSSD activity, or (3) a response action for a major fiber release episode. All persons seeking accreditation as asbestos abatement contractor/supervisors shall complete at least a 5–day training course as outlined below. The training course must include lectures, demonstrations, at least 14 hours of hands-on training, individual respirator fit testing, course review, and a written examination. Hands-on training must permit supervisors to have actual experience performing tasks associated with asbestos abatement.
40 CFR Part 763, Subpart E, App. C., Unit III.B

Asbestos Worker – Initial

EPA Model Accreditation Program: A person must be accredited as a worker to carry out any of the following activities with respect to friable ACBM in a school or public and commercial building: (1) A response action other than a SSSD activity, (2) a maintenance activity that disturbs friable ACBM other than a SSSD activity, or (3) a response action for a major fiber release episode. All persons seeking accreditation as asbestos abatement workers shall complete at least a 4–day training course as outlined below. The 4–day worker training course shall include lectures, demonstrations, at least 14 hours of hands-on training, individual respirator fit testing, course review, and an examination. Hands-on training must permit workers to have actual experience performing tasks associated with asbestos abatement. A person who is otherwise accredited as a contractor/supervisor may perform in the role of a worker without possessing separate accreditation as a worker.
40 CFR Part 763, Subpart E, App. C., Unit III.B

OSHA Construction Standard: For Class I and II asbestos work the competent person shall be trained in all aspects of asbestos removal and handling, including: abatement, installation, removal and handling; the contents of this standard; the identification of asbestos; removal procedures, where appropriate; and other practices for reducing the hazard. Such training shall be obtained in a comprehensive course for supervisors that meets the criteria of EPA's Model Accredited Plan (40 CFR part 763, subpart E, Appendix C), such as a course conducted by an EPA-approved or state-approved training provider, certified by EPA or a state, or a course equivalent in stringency, content, and length.

- ***Class I asbestos work*** means activities involving the removal of TSI and surfacing ACM and PACM.
- ***Class II asbestos work*** means activities involving the removal of ACM which is not thermal system insulation or surfacing material. This includes, but is not limited to, the removal of asbestos-containing wallboard, floor tile and sheeting, roofing and siding shingles, and construction mastics.
- 29 CFR 1926.1101(o)(4)

Asbestos Contractor / Supervisor – Refresher
EPA Model Accreditation Program: A State's accreditation program shall include annual refresher training as a requirement for reaccreditation as indicated below:
- Contractor/Supervisors: One full day of refresher training.
40 CFR Part 763, Subpart E, App. C., Unit III.B

Asbestos Contractor / Supervisor – Refresher
EPA Model Accreditation Program: A State's accreditation program shall include annual refresher training as a requirement for reaccreditation as indicated below:
- Contractor/Supervisors: One full day of refresher training.
40 CFR Part 763, Subpart E, App. C., Unit III.B

This manual was developed to provide guidance for those who are about to undertake an asbestos abatement project. The procedures and practices detailed in these pages incorporate current technology at the time of publishing. The reader should be reminded that as technology evolves, so do the methods for conducting asbestos abatement. Therefore, we emphasize the need for the reader to obtain the most up-to-date information available.

Many of the recommended procedures discussed in this document go beyond the minimum requirements of the various regulations that have been promulgated by the U.S. Occupational Safety and Health Administration (OSHA) and the U.S. Environmental Protection Agency (EPA). Recommended practices are not mandatory requirements. The reader should refer to these regulations to determine federal requirements for asbestos abatement projects: OSHA Asbestos Standard for the Construction Industry (29 CFR 1926.1101), and EPA's National Emission Standards for Hazardous Air Pollutants (40 CFR 61, subpart M), Asbestos Hazard Emergency Response Act (40 CFR Part 763), and Asbestos School Hazard Abatement Reauthorization. Also, the reader should be aware of existing additional procedures that may be substituted for those discussed here. Further, the practices recommended may not be appropriate for every project.

This manual was prepared by professionals in the field of asbestos abatement and control and revised (ehsMaterials.com) as of September 2012. Throughout its preparation and upon completion, the manuscript was subjected to review in government, academic, and industry circles.

The technical expertise and common sense provided by the contractor are major components of a successful abatement project. We encourage the reader to improve further upon the techniques provided in this manual as he/she gains additional knowledge through field practice. This will ensure that the abatement industry continues to evolve to improve asbestos abatement work practices.

State regulations vary and can be more stringent than Federal requirements.

Chapter 1 – COURSE OVERVIEW

Objective: To provide a brief discussion of the topics that will be covered and how these topics are integral components of an asbestos abatement project; acquaint participants with the types of questions on the post-course exam.

Learning Tasks: Information in this chapter should enable participants to:

- Become familiar with the contents of the manual
- Learn what topics will be covered and in what sequence
- Become aware of the various facets of an asbestos abatement project
- Be introduced to a multidisciplinary approach to asbestos abatement
- Perform a self-evaluation of their knowledge concerning the subject via a pre-course quiz
- Become familiar with the format and contents of the post-course exam

This overview of the training program for asbestos abatement contractors and supervisors demonstrates how different parts of the course fit together. Since contractors are often consulted as asbestos specialists, it will also assist them in addressing the asbestos abatement issue with the scientific community and general public.

The public is often confused about the source of asbestos. It is mined from the earth in much the same ways as iron, lead, and copper. Instead of breaking up into dust particles, asbestos divides into millions of fibers, some of which may be so fine as to be undetectable to the naked eye. Asbestos fibers include the asbestiform varieties of: chrysotile (serpentine); crocidolite (riebeckite); amosite (cummingtonite; grunerite); anthophyllite; tremolite; and actinolite. All varieties exhibit substantial resistance to heat and chemicals and thus have been used for a variety of commercial and industrial purposes. In fact, asbestos has been used in more than 3,000 products.

The asbestos industry started during the 1870s when the first commercial chrysotile mine opened in Quebec, Canada. The crocidolite variety was first mined in South Africa during the 1890s. Amosite also comes from Africa, but its mining did not begin until 1916.

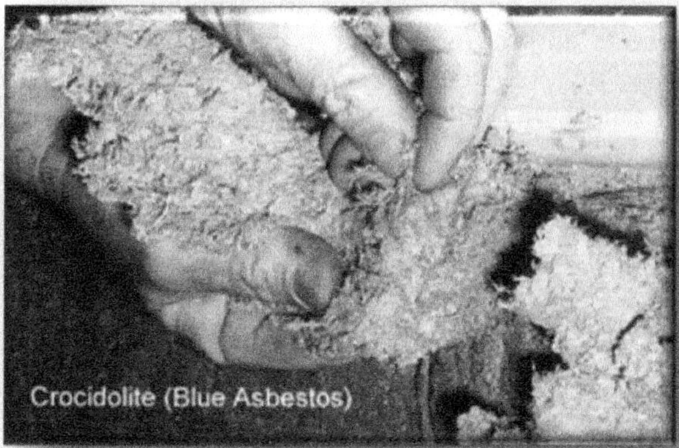

Crocidolite (Blue Asbestos)

Asbestos was used in limited amounts as insulation during the period 1870-1900, but its use became more common on steam pipes and boilers of ships after 1900. Until the early 1940s, most asbestos-containing insulations in the United States contained chrysotile from Canada.

The 1940s marked significant changes in the American use of asbestos. Amosite became widely used in American ships and shipyards during World War II. The use of asbestos then started expanding to include acoustical and decorative purposes, especially in buildings. After WWII ended and military demand for asbestos declined, sprayed asbestos fireproofing materials were used extensively in buildings. Estimates indicate that more than half of the large multi-story buildings constructed during the 1950-1970 period contain some form of sprayed asbestos-containing materials. These uses will be discussed and shown in the next chapter, "Background Information on Asbestos." Asbestos use in the United States didn't start to decline until the 1973-1978 EPA bans on spray-applied materials and the building recession, which happened during that same period.

Diseases and deaths associated with exposure to asbestos fibers are the principle factors behind "The Asbestos Problem." The three major respiratory diseases associated with asbestos are asbestosis, lung cancer, and mesothelioma. Their medical natures are discussed in much greater detail in the chapter "Health Effects of Asbestos Exposure."

While the documented health effects from exposure to asbestos are no doubt the underlying cause of the asbestos problem, the legal problems that followed and even now continue are contributing factors. Legal actions affect all parties involved in asbestos abatement in several ways and are addressed in the chapter "Legal and Insurance Considerations."

During the early 1970s, the federal government attempted to respond to these health and legal problems by issuing regulations and guidelines to provide some controls for industrial exposures to asbestos fibers. Then during the late 1970s and early 1980s, they added guidelines and requirements to cover building occupants, especially schools. These actions, combined with the health and legal issues, prompted the need for asbestos abatement policies.

Before an asbestos problem in a building can be controlled, it must be evaluated and a variety of decisions must be made. Those evaluating and deciding techniques are not the focus of this course, but since they can affect all involved parties, they will be reviewed during the next chapter of this manual.

If an asbestos problem exists, the evaluation and decision techniques will almost always result in "Contract Specifications" for an abatement or control project. Those specifications will be discussed in a chapter devoted to that subject. Once the specifications are issued and a contractor is selected to do the work, a variety of simultaneous and sequential considerations, techniques, and job phases are necessary to complete the abatement project. These considerations, techniques, and phases are the main contents of this course.

PRE-COURSE QUIZ

1 - Asbestos

A. is a natural occurring mineral
B. is a synthetic product developed within the last twenty years
C. was never used until the 1970's
D. was only used in schools

2 - A substance when dry can easily be crumbled is called

A. pliable
B. fibrous
C. friable
D. durable

3 - Asbestos has the following unique properties;

A. fire resistance
B. impervious to some chemicals
C. high tensile strength
D. all of the above

4 - Asbestosis

A. is a kind of cancer
B. results in fibrotic scarring of the lung tissue
C. may affect the joints
D. may cause blindness

5 - Asbestos related diseases include all, EXCEPT

A. asbestosis
B. heart disease
C. lung cancer
D. mesothelioma

6 - An asbestos worker, who smokes, is ______ many times more likely to be diagnosis with lung cancer

A. 5
B. 10
C. 50
D. there is no difference between smoker and non-smoker

7 - During the decontamination process, in which room should the worker remove their suit?

A. equipment room
B. clean room
C. air-lock
D. shower

8 - The period it takes the disease to manifest itself is called

A. duration
B. linear
C. latency
D. potency

9 - Which type of respirator is NOT allowed in the asbestos abatement industry

A. Type C air supplied
B. Dust mask
C. Half-face air purifying
D. Full-face powered air purifying

10 - The type of filters that are used in the asbestos abatement industry are called

A. HETA
B. HEPA
C. MSHA
D. NIOSH

Objective: To become familiar with the history and uses of asbestos.

Learning Tasks: Information in this chapter should enable participants to:

- Recognize the characteristics and various types of asbestos.
- Become acquainted with various types of asbestos-containing materials (ACM) found in building applications.

HISTORICAL PERSPECTIVE

The word asbestos is derived from a Greek adjective meaning **inextinguishable**. The "**miracle mineral,**" as it was referred to by the Greeks, was admired for its soft and pliant properties, as well as its ability to withstand heat. Asbestos was spun and woven into cloth in the same manner as cotton. It was also utilized for wicks in sacred lamps. Romans likewise recognized the properties of asbestos, and it is thought that they cleaned asbestos tablecloths by throwing them into the flames of a fire.

From the time of the Greeks and Romans in the first century until its re-emergence in the eighteenth century, asbestos received little attention or use. It was not available in large amounts until extensive deposits were discovered in Canada in the nineteenth century (late 1800s). Following this discovery, asbestos emerged as an insulating component in thermal insulation for boilers, pipes, and other high temperature applications and as a reinforcement material for a variety of products.

CHARACTERISTICS OF ASBESTOS

Asbestos is a naturally occurring mineral. It is distinguished from other minerals by the fact that its crystals form long, thin fibers. Deposits of asbestos are found throughout the world. The primary sites of commercial production are: the Commonwealth of Independent States, Canada, China, Brazil, Zimbabwe, and South Africa. Asbestos is also mined commercially in limited quantities in the United States, in California, and Vermont.

Asbestos minerals are divided into two groups – **serpentine and amphibole**. The distinction between groups is based upon a mineral's crystalline structure – serpentine minerals have a sheet or layered structure, amphiboles have a chain- like crystal structure. Chrysotile, the only asbestos mineral in the serpentine group, is the most commonly used type of asbestos and accounts for approximately 95% of the asbestos found in buildings in the United States. Chrysotile is commonly known as **"white asbestos,"** so named for its natural color.

Five types of asbestos are found in the amphibole group. **Amosite,** the second most likely type to be found in buildings, is often referred to as **"brown asbestos."** As you might assume, in its natural state, amosite is brown in color.

Crocidolite, "blue asbestos," is also an amphibole. Crocidolite was used in high temperature insulation applications.

The remaining three types of asbestos in the amphibole group are: **anthophyllite, tremolite, and actinolite**. These varieties are of little commercial value. Occasionally, they are found as contaminants in asbestos- containing materials.

Once extracted from the earth, asbestos-containing rock is crushed, milled (ground), and graded. This produces long, threadlike fibers of material. What actually appears as a fiber is an agglomeration of hundreds or thousands of fibers, each of which can be divided even further into microscopic fibrils

USES OF ASBESTOS

Asbestos has been used in literally thousands of products. Collectively, these are frequently referred to as asbestos-containing material (ACM). Asbestos gained widespread use because it is plentiful, readily available, and low in cost. Because of its unique properties – fire resistance, high tensile strength, poor heat electrical conductivity, and being generally impervious to chemical attacks – asbestos proved well suited for many uses in the construction trades.

One of the most common uses for asbestos has been as a fireproofing material. It was sprayed on steel beams, columns and decking that were used in construction of multi-storied buildings. This application prevented these structural members from warping or collapsing in the event of fire.

Chrysotile was the most commonly used asbestos constituent in sprayed-on fireproofing. Asbestos comprised 5 – 95 percent of the fireproofing mixture and was used in conjunction with materials such as vermiculite, sand, cellulose fibers, gypsum, and a binder such as calcium carbonate. These materials are soft and may be fluffy in appearance and to the touch. They vary in color from white to dark gray; occasionally they have been painted or encapsulated with a clear or colored sealant. The material may be exposed or concealed behind a suspended ceiling. Application to structural members (beams and columns) often resulted in some material being sprayed on walls and ceilings as well. This is referred to as overspray.

Asbestos is added to a variety of building materials to enhance strength. It is found in concrete and concrete-like products. Asbestos-containing cement products generally contain Portland cement, aggregate, and chrysotile fibers. The asbestos content may range up to 50 percent by weight depending on the use of the product. Asbestos cement products are used as siding and roofing shingles; as wallboard; as corrugated and flat sheets for roofing, cladding, and partitions; and as pipes. Asbestos has also been added to asphalt, vinyl, and other materials to make products like roofing felts, exterior siding, floor tile, joint compounds and adhesives. A report issued by the U.S. Bureau of Mines in 1993 indicated that approximately 70% of the asbestos used in the U.S. in 1992 was incorporated into asbestos cement pipe, coatings, compounds, packings, and roofing products. Fibers in asbestos cement, asphalt, and vinyl are usually firmly bound in the cement and will be released only if the material is mechanically damaged, for example, by sanding, grinding, cutting, or abrading. Roofing shingles and siding may also show slow deterioration due to weathering.

As an insulator, asbestos received widespread use for thermal insulation and condensation control. It was usually spray applied, trowel applied, or factory installed on or within equipment.

Asbestos proved valuable as a component of acoustical plaster. The material was applied by trowel or by spraying on ceilings and sometimes walls. It varies in color from white to gray – rarely was it painted, as a noticeable loss of acoustical value occurs. As a decorative product, asbestos was mixed with other materials and sprayed on ceilings and walls to produce a soft, textured appearance.

FRIABLE vs NONFRIABLE ACM

The U.S. Environmental Protection Agency (EPA) distinguishes between friable and nonfriable forms of ACM. Friable ACM contains more than 1% asbestos and can be "crumbled, pulverized, or reduced to powder by hand pressure when dry." Other things being equal, friable ACM is thought to release fibers into the air more readily; however, many types of nonfriable ACM can also release fibers if disturbed.

FRIABLE means easily crumbled into dust

NON-FRIABLE means difficult to crumble into dust

CATEGORIES OF ASBESTOS-CONTAINING BUILDING MATERIALS

EPA identifies three categories of ACM used in buildings

Surfacing Materials – ACM sprayed or troweled on surfaces (walls, ceilings, structural members) for acoustical, decorative, or fireproofing purposes. This includes plaster and fireproofing insulation.

Thermal System Insulation – Insulation used to inhibit heat transfer or prevent condensation on pipes, boilers, tanks, ducts, and various other components of hot and cold water systems and heating, ventilation, and air conditioning (HVAC) systems. This includes pipe lagging; pipe wrap; block, batt, and blanket insulation; cements and "muds;" and a variety of other products such as gaskets and ropes.

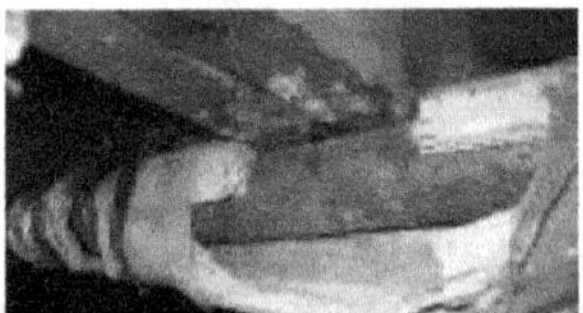

Miscellaneous Materials – Other, largely non-friable products and materials such as floor tile, ceiling tile, roofing felt, concrete pipe, outdoor siding, and fabrics.

While it is often possible to “suspect” that a material or product is or contains asbestos by visual determination, actual determinations can only be made by instrumental analysis. The EPA requires that the asbestos content of suspect materials be determined by collecting bulk samples and analyzing them by polarized light microscopy (PLM). The PLM technique determines both the percent and type of asbestos in the bulk material.

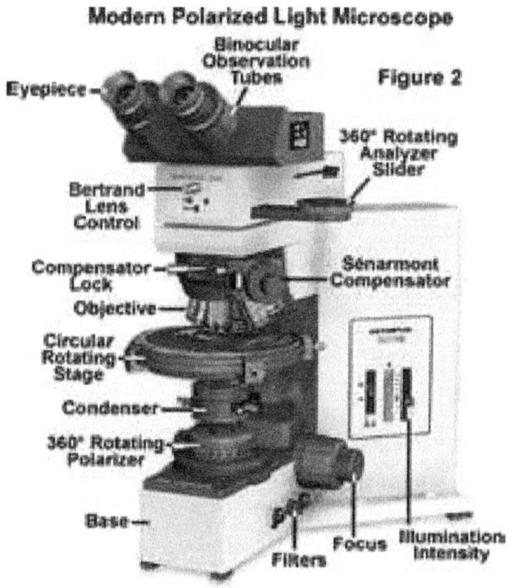

Objective: To provide a brief overview of the mechanisms of exposure and diseases associated with inhalation of asbestos fibers

Learning Tasks: Information in this chapter should enable participants to:

- Gain a brief understanding of the means by which asbestos can enter the body and cause damage.
- Conceptually understand the major diseases associated with asbestos exposure.
- Understand the concept of latency period, or length of time following exposure to asbestos before the onset of disease.
- Understand the relationship between cigarette smoking and asbestos exposure, and the increased risk of disease.
- Gain a brief overview of the risks associated with asbestos exposure, or how likely disease will occur.
- Recognize the need for medical surveillance and when it is necessary.

The adverse health effects associated with asbestos exposure have been extensively studied for many years. Results of these studies and epidemiologic investigations have demonstrated that inhalation of asbestos fibers may lead to increased risk of developing one or more diseases. Exactly why some people develop these diseases and others do not remain a mystery. In this discussion, each of the major diseases associated with asbestos will be examined, along with the risk and how that risk can be minimized.

It is important to recognize that the majority of people who have died as a result of asbestos exposure were workers employed in the mining, milling, manufacturing, and insulating industries, who worked with raw or processed asbestos. These workers were frequently exposed to high concentrations of asbestos fibers each working day with little or no protection.

The asbestos abatement worker of today follows specific work practices and wears appropriate protection, including respirators, to minimize the risk of exposure.

THE RESPIRATORY SYSTEM

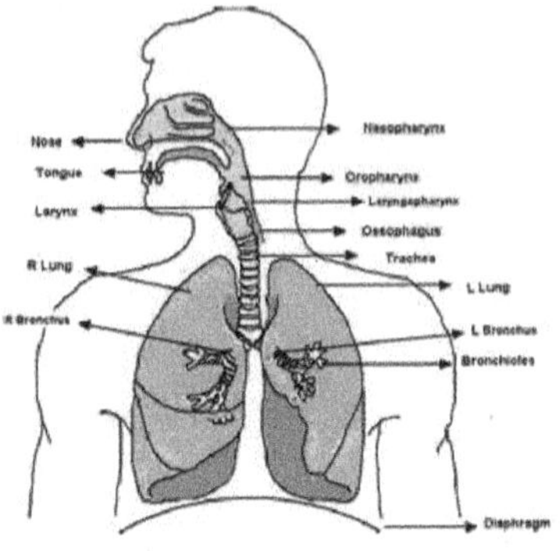

Since the primary health effects due to asbestos exposure act on the lung, it is necessary to gain a brief understanding of the respiratory system. Air is breathed into the body and passes through the mouth and nose into the windpipe or trachea. The trachea splits into two smaller airways called the bronchi. Each bronchus divides into smaller and smaller tubes that terminate into air sacs called alveoli. It is in these air sacs that oxygen is absorbed into small blood vessels, and waste gases, such as carbon dioxide, pass out of the blood.

The lung itself is divided into two halves and sits in the pleural cavity.This cavity and the outside of the lung itself have a membrane lining (called the pleura) which looks somewhat like plastic food wrap. These linings are in contact with each other and are very moist. Just like two panes of glass with a drop of water between them, these linings slide easily across each other, but are difficult to pull apart. Accordingly, as the chest cavity expands, the lungs expand and air rushes in. If these linings were to become damaged, inhalation could not occur properly.

The body has several mechanisms by which it "filters" the air it breathes. First, very large particles are removed in the nose and mouth. Many smaller particles impact on the mucous-coated walls of the airways and are caught. These airways have a hair-like lining (ciliated cells) that constantly beats upward. Accordingly, particles caught in the mucous are swept up into the back of the mouth. From here they are swallowed or expelled. Unfortunately, cigarette smoking temporarily paralyzes these ciliated cells, inhibiting one of the body's natural defenses against unwanted dust. Poorly functioning cilia result in secretions pooling in the lungs. During sleep, in the absence of smoke, the hair-like cells start working again and carry large amounts of mucous into the back of the mouth. This causes the so-called "smoker's hack" upon awakening.

After the first cigarette or two, the cleansing mechanism is paralyzed again and the coughing stops. It should now be evident why cigarette smokers who are exposed to asbestos appear to be at greater risk of developing lung cancer and possibly other asbestos-related diseases. Other reasons will also be discussed later in this chapter.

Mesothelial cells line the outside of the lungs and the pleural and peritoneal cavities. Interaction of asbestos with these cells can result in either pleural or peritoneal mesothelioma.

Even with the above-mentioned natural defenses of the body, some dust particles inevitably reach the tiny air sacs. When this occurs, large cells (called macrophages) attempt to engulf the particle and "digest" it. For this reason, they are sometimes called the lung's garbage collectors. However, because of chemical and physical properties of the asbestos fiber, the macrophages are often not able to digest the fiber. In the process of trying to ingest the asbestos particle, the macrophage is frequently damaged or destroyed causing caustic enzymes to be released into sensitive lung tissue. Macrophages also emit signals to other "digestive" cells which also release enzymes. This action can turn into a chronic process resulting in scar formation and a walling off of the asbestos fiber. If many asbestos fibers are inhaled and much scar tissue is formed, a condition develops, known as asbestosis.

ASBESTOSIS

Asbestosis is a disease characterized by fibrotic scarring of the lung. This is a restrictive lung disease that reduces the overall volume of the lung. The common symptom is shortness of breath. Asbestosis is prevalent among workers who have been exposed to large doses of asbestos fibers over a long period of time. Accordingly, there is a clear dose-response relationship between asbestos exposure and developing this disease. This means the greater the asbestos exposure, the more likely asbestosis will develop. All forms of asbestos have demonstrated the ability to cause asbestosis. Like all diseases associated with asbestos exposure, it may take many years for the disease to show up. The typical latency period for asbestosis is 10-20 years.

Even after exposure to asbestos has ceased, scar tissue will continue to form around existing scar tissue and fibers in the lung. Limiting exposure will reduce the amount of new scar tissue since additional fibers entering the lung will be reduced. The current Occupational Safety and Health Administration (OSHA) Asbestos Standards (29 CFR 1910.1001 and 29 CFR 1926.1101) were promulgated to greatly reduce asbestosis among asbestos workers by reducing their daily dose of asbestos.

LUNG CANCER

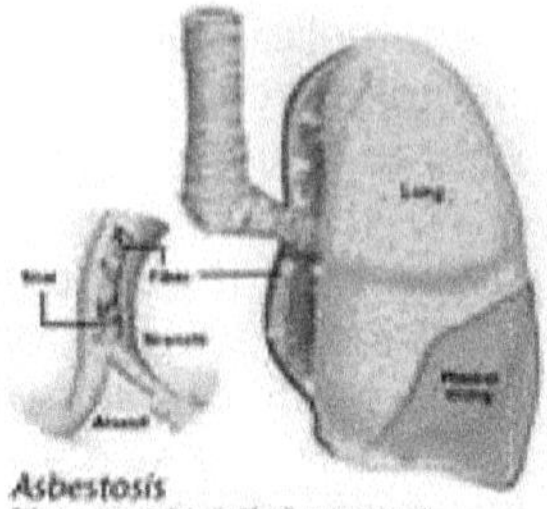

Asbestosis
Fibres accumulate in the lungs narrow branches, inflaming and scarring airways. The condition causes chronic cough and chest pain.

There are many causes of lung cancer, of which asbestos is only one. While employees exposed to industrial concentrations of asbestos in years past have an increased risk of getting lung cancer (five times greater than the general population), their risk is not as great as the cigarette smoker (ten times greater than the general population). However, together, a cigarette smoker who also works with asbestos is more than 50 times more likely to contract lung cancer than the normal non-smoking population. This relationship between asbestos exposure and cigarette smoking is termed a synergistic relationship. As with asbestosis, there exists a long lag time between initial exposure and the occurrence of lung cancer, typically 20 years. There appears to be a dose- response relationship between asbestos exposure and lung cancer, although no "safe level" has yet been determined. It should be noted, however, that several research papers published in the mid-1980s suggest there may exist an exposure level for certain forms of asbestos below which the occurrence of lung cancer related to asbestos exposure will not exceed that of the general population.

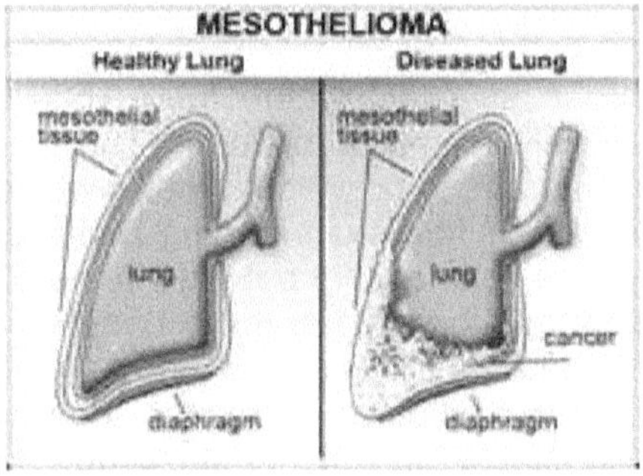

The asbestos-associated disease of greatest concern regarding asbestos in buildings is probably mesothelioma. Fortunately, it is also the rarest.

MESOTHELIOMA - Mesothelioma is a cancer of the chest cavity lining (mesothelium). Mesothelioma can also occur in the lining of the abdominal cavity. If it occurs in the chest cavity, it is called pleura mesothelioma; in the abdominal cavity, it is known as peritoneal mesothelioma. This type of cancer spreads very rapidly and is always fatal. The exact mechanism of this disease remains unknown. There does not appear to be any increased risk of getting mesothelioma for smokers, nor does there appear to be a dose-response relationship between asbestos exposure and mesothelioma. Studies indicate that **crocidolite** asbestos exposure is more closely linked to mesothelioma than the other types of asbestos. However, it is very uncommon for an asbestos-exposed individual to be exposed to only one of the asbestos minerals. Cases have been recorded where the person's asbestos exposure has been limited, such as Steve McQueen, the actor. Similar to the other diseases of asbestos, mesotheliomas have a long latency period and may not develop for 20 to 40 years after initial exposure.

OTHER DISEASES

Several other diseases are found more often among persons exposed to asbestos than the normal population. These include cancer of the esophagus, stomach, colon, and pancreas, pleural plaques, pleural thickening, and pleural effusion. Again, the importance of using the proper work practices and respiratory protection cannot be overemphasized to minimize the occurrence of these diseases due to unnecessary asbestos exposure.

RISKS ASSOCIATED WITH LOW LEVEL ASBESTOS EXPOSURE

Asbestos is known to be hazardous based on studies of asbestos workers and laboratory animals. However, the risks associated with low level, non-occupational exposure (for example, as an occupant of a building containing ACM) are not well established. Attempts have been made to estimate low level risks by extrapolation from occupational exposure data. This is not a straightforward process and its validity is questionable.

A 1988 survey sponsored by EPA attempted to assess exposure to ACM in public and commercial buildings. According to the data, a lower percentage of public and commercial buildings contain friable ACM than do school buildings (20% vs. 35%). However, limitations in the data prevent firm conclusions regarding the number of persons exposed, exposure levels, or the exposure levels of service/maintenance workers in comparison with the public.

Additional studies on low-level asbestos exposures are ongoing. According to a literature review conducted by the Health Effects Institute-Asbestos Research, it is theorized, with constraints, that if asbestos workers were exposed to a level of 0.1 f/cc for 20 years, the risk of premature cancer deaths (lifetime risks) would be 2 in 1,000. Using the same theoretical calculations, it is estimated that, based on the average asbestos level (0.0002 f/cc) for the buildings reviewed in their survey, for occupants of these buildings the risk of premature cancer deaths would be 4 per 1,000,000.

A mathematical model has been developed by EPA to assess risk. Risk calculations suggest that if asbestos exposure is eliminated in schools, we have the potential to significantly reduce the overall risk for the segment of our population that may later be exposed to asbestos in public and commercial buildings. It should be noted, however, that although the elimination of exposure in schools may reduce risk, there remains a risk as a result of exposure to asbestos elsewhere.

Asbestos fibers accumulate in the lungs. As exposure increases, the risk of disease likewise increases. Measures to minimize exposure, and consequently minimize the accumulation of fibers, reduce the risk of adverse health effects.

Based on a thorough review of the health effects literature, EPA concludes that there is no level of exposure under which the risks of contracting an asbestos-related disease are zero. That is, there is no zero threshold level of exposure. Despite the uncertainties associated with the risk of low level exposure, if we accept the fact that there is no safe level of exposure to asbestos, we have cause to institute measures to control or eliminate exposure; regulations such as AHERA move in this direction.

MANUFACTURED MINERAL FIBERS

Many of the non-asbestos substitutes for asbestos in buildings are manufactured mineral fibers. This is a generic term applied to fibrous inorganic material made primarily from rock, clay, slag, or glass. Recently, concern has grown regarding the potential health effects resulting from exposure to these fibers. It is generally accepted that many of these replacement insulation products have a thicker fiber diameter, which makes them less respirable than the very thin fibers of the asbestos minerals. However, as manufacturing and extruding processes advance, some of these manufactured mineral fibers are achieving thinner, more respirable diameters. Studies indicate that refractory ceramic fibers, which are thin, respirable, and durable, may be more of a potential cancer risk than previously believed.

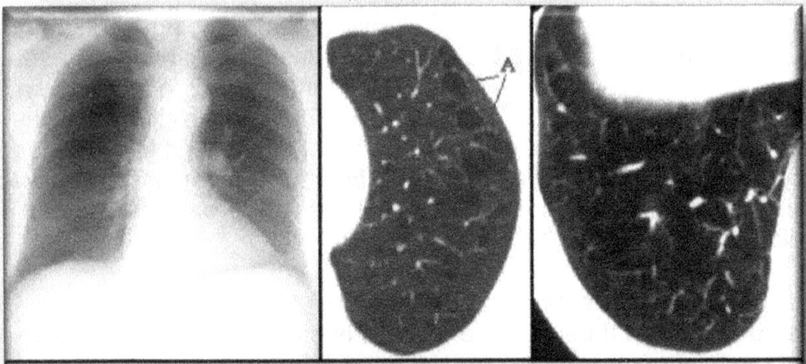

50-year-old male asbestos-exposed cigarette smoker. The chest-x ray was read as possible asbestosis. HRCT scan (middle) through the upper lobes shows changes of emphema without fibrosis (A). The same findings are present in the lower lobes.

Objectives: To provide a detailed discussion of the use, maintenance, and limitations of respiratory protection and protective clothing.

Learning Tasks: Information in this chapter should enable participants to:

- Identify the need for effective respiratory protection for asbestos abatement personnel.
- Understand the categories and operating principles of respirators used for protection against asbestos.
- Recognize the use and limitations of various types of respirators.
- Understand the important of properly fitting the respirator.
- Become familiar with the concept of protection factors and how they relate to respirator selection and use.
- Understand the basic requirements of an effective respiratory protection program.
- Recognize the need for and proper use of protective clothing and equipment.

INTRODUCTION

When asbestos-containing materials are disturbed, as in the case of asbestos repair, encapsulation, enclosure, or removal projects, asbestos fibers will become airborne. Once in the air, the asbestos fibers may be inhaled by workers performing these projects, posing a significant health risk. For these reasons, engineering controls and work practices such as wet methods are used to minimize the generation of airborne fibers. Since no method has been devised to remove asbestos without generating some airborne fibers, respirators are used. Further, protective clothing is worn. Lastly, other protective measures such as safety glasses, hard hats, or ear plugs may be required to protect the employee from other hazards.

There are three ways that hazardous materials can enter the body:

(1) through the gastrointestinal tract, usually via the mouth;
(2) through the skin; and
(3) through the respiratory systems.

Asbestos does not appear to pose a serious threat to the body through the second routes of entry. It can, however, cause serious diseases when it enters the body primarily through the respiratory system, and secondarily through the gastrointestinal tract.

RESPIRATORY SYSTEM

The respiratory system is an air pump containing a series of airways leading from the nose and mouth down into the air sacs (alveoli) of the lung where there is an exchange of oxygen and carbon dioxide. The main components of the respiratory system, from top to bottom are as follows:

- Nose and mouth
- Throat
- Larynx (voice box)
- Trachea (windpipe)
- Bronchi (branches from trachea)
- Alveoli (air sacs in the lung)
- Diaphragm and chest muscles

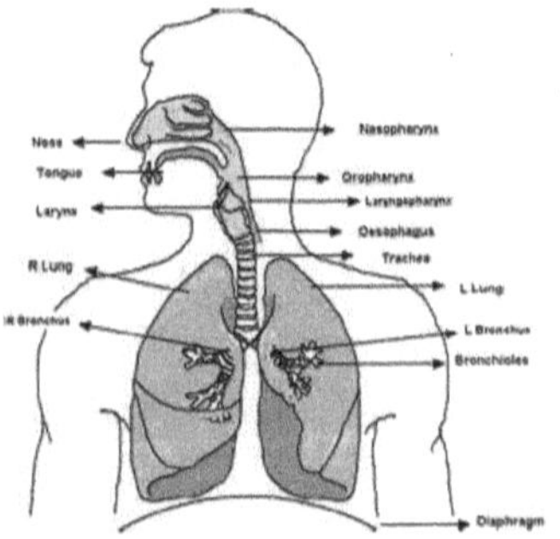

The human body has certain natural defenses to protect itself against inhaling dust, the most important being the muco-ciliary escalator. Airways of the upper respiratory tract (trachea through bronchi) are lined with cilia (hair-like protrusions) covered with a layer of mucous. These cilia are constantly sweeping upward quickly, then down slowly, and thus moving the mucous and trapped materials up at a rate of approximately one inch per minute. This is an important clearance mechanism that prevents most large particles from reaching the alveoli in the lungs. Particles trapped in the mucous are carried back up to the throat where they are swallowed or expectorated.

Unfortunately, this natural defense mechanism does not prevent all asbestos fibers from reaching the lung where damage can occur. Accordingly, respirators are worn to provide further protection when asbestos exposure is likely.

RESPIRATORY HAZARDS

Respiratory hazards are generally divided into two categories:

- Oxygen deficiency and
- Toxic contaminants.

Generally, asbestos abatement projects do not pose oxygen deficiency hazards. However, since there may be rare projects and circumstances where it can be a problem, oxygen deficiency must always be considered. For example, there could be an oxygen deficiency problem while performing abatement in steam tunnels, mechanical chases, or boilers. Failing to consider oxygen deficiency could result in a fatality on any project.

Toxic contaminants are a more common category of respiratory hazards encountered on abatement projects. These toxic contaminants are generally divided into three categories:

- Particulates;
- Gases and vapors; and
- Any combination of the above.

Asbestos fibers are an example of the particulate category, carbon monoxide is an example of the gaseous category, and an epoxy encapsulant is an example of a harmful organic vapor. It is possible to have all of these hazardous substances, as well as others, in a work area at the same time.

The control of respiratory hazards often involves three steps:

- Assessing the hazards;
- Reducing or eliminating the hazards; and
- Providing respiratory protective equipment.

The asbestos detection and control industry is actually based on these first two steps. Buildings and other structures are inspected or surveyed to assess potential asbestos hazards. When a potential asbestos hazard exists, a contractor is called upon to reduce or eliminate the hazard through removal, encapsulation, or enclosure of the material. Thus, the third step, the use of respirators to protect the building occupants, custodial, and maintenance personnel, can be minimized.

In the case of worker protection, the same three steps can be followed. Hazard assessment usually means identifying potential air contaminants and evaluating their concentrations through air sampling of worker breathing zones. Reduction of airborne hazards can be accomplished through cautious work practices such as wetting work surfaces, the use of glove bags, and negative air enclosures. Finally, the appropriate respiratory protection is provided based upon the environmental conditions, work activities, and the needs of the workers.

CATEGORIES OF RESPIRATORS

There are four broad categories of respirators. These are:

(1) Air-purifying respirators;
(2) Supplied-air respirators;
(3) Self-contained breathing apparatus (SCBA); and
(4) Combination respirators.

In each category there are many different types of respirators (i.e., powered air-purifying respirators, gas masks, pressure demand supplied-air respirators, etc.). Many of the respirators available for use, however, are not appropriate for protection against asbestos.

Air-Purifying Respirators

These respirators remove the hazardous contaminant from the breathing air before it is inhaled. They consist of a soft rubber or synthetic elastomeric face piece and replaceable filters or cartridges.

Because they depend upon the breathing action of the wearer to draw air through the respirator filter or cartridge where it is decontaminated, air-purifying respirators that rely on a person's lung power are referred to as "negative pressure" respirators.

Two major subcategories of air-purifying respirators are the "particulate-removing" mechanical filter type and the "gas- and vapor-removing" chemical cartridge type. The mechanical filter variety is designed to protect against particulate contaminants such as asbestos. The chemical cartridge type protects against gaseous contaminants such as solvent vapors. Because each respirator assembly is approved for a particular contaminant, care must be taken in choosing the appropriate unit. Protection from asbestos requires high efficiency particulate air (HEPA) filters. The HEPA filters used for asbestos must be 99.97 percent efficient at removing particles 0.3 µm (micrometers) in diameter.

Filter cartridges that provide protection against other air contaminants are usually color-coded accordingly; some units have both particulate- and gas/vapor-removing capabilities and have dual color codes and are typically purple or magenta in color.

Air-purifying respirators are further categorized based on their degree of face coverage. The half-mask respirator covers half the face – from the bridge of the nose to under the chin. A full-face respirator covers the face from the forehead to under the chin, providing a better fit and a higher degree of protection than the half-mask.

Powered Air Purifying Respirator (PAPR)

A special subcategory of an air-purifying respirator is the Powered Air Purifying Respirator(PAPR). These respirators also require filters with an efficiency rating of 99.97 percent efficient at removing particles 0.3 µm (micrometers) in diameter. PAPRs, however, are positive-pressure devices that employ a portable, rechargeable battery pack and blower to force contaminated air through a filter or cartridge, where it is cleaned and supplied to the wearer's breathing zone. PAPRs designed for asbestos use are tight-fitting and usually full-face piece. An advantage of using a powered air-purifying respirator is that it supplies air at a positive pressure within the face piece and leakage occurs in an outward direction. Because the air is being drawn from the immediate work area, they, like other air-purifying respirators, offer no protection against oxygen deficient atmospheres.

Air purifying respirators remove limited concentrations of air contaminants from the breathing air, but do nothing to improve (or change) the oxygen content or conditions where air contaminants exceed the specified range of the respirator and cartridge. Often, however, this is adequate protection when prepping the asbestos abatement work area, performing final cleanup (wipe downs), or during glovebag removal projects. But during gross removal and gross cleanup, a different category of respirator is used to provide greater protection for the worker: the supplied-air respirator.

Supplied-Air Respirators

These respirators supply breathing air from a source independent of the work area atmosphere. Air is delivered to the face piece through an airline(hose). These are referred to as "airline" or "Type C" respirators. There are three types of supplied-air respirators, based upon how the airflow to the face piece is regulated:

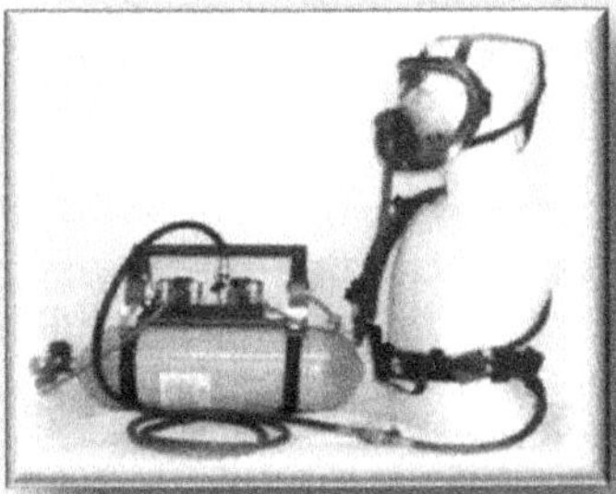

- Constant or Continuous Flow
- Demand
- Pressure Demand

In the case of constant flow devices, air is being continuously supplied to the face piece. Although air is constantly supplied to the wearer, the continuous flow respirator is not considered to a positive pressure respirator. Depending on the physical condition of the worker, the demand from breathing can exceed the rate air is supplied. The continuous flow cannot increase the air flow rate to the wearer during these times, resulting in a negative pressure inside the mask.

Demand regulators provide air to the wearer only during inhalation. This means that the wearer must first create a negative pressure zone inside the face piece (with respect to outside the face piece) before the regulator supplies the air. Under this condition, should leakage occur, air will move into the face piece from the contaminated work environment. Demand and continuous flow regulators are classified as negative-pressure respirators and are generally not recommended for use in situations where the higher protection afforded by Type C respirators is required.

A pressure-demand regulator always maintains a positive pressure inside the face piece and supplied-air devices so equipped are considered the most protective.

A continuous supply of air is provided to the user, and additional doses of air are supplied when the wearer inhales. In the event of leakage around the face piece-to-face seal, air will move from inside the face piece in an outward direction.

Because of the importance and complexity of supplied-air respirators, the entire next chapter of this manual is devoted to the topic of "Type C" supplied-air systems.

Combination Respirators

Supplied-air respirators equipped with either an emergency HEPA filter(s), or 5-or 10-minute SCBA, are designated as "Combination Respirators."In the event of failure of the supplied-air component of a combination respirator, the wearer is afforded a means of safely leaving the work environment to a safe area, and the air supply can be subsequently restored. It should be noted that neither the air-purifying element (HEPA filter) nor the escape SCBA should be used routinely. Once used, the emergency HEPA filter should be discarded and replaced, and the escape SCBA cylinder should be immediately refilled.

RESPIRATORY PROTECTION PROGRAM

Although respirators are commonly used to help protect against inhalation hazards, having a respiratory protection program does not mean simply donning a respirator and expecting to be adequately protected. Instead, a comprehensive program is necessary to ensure proper protection including proper equipment selection, effective worker training, regular medical evaluation and fit-testing of workers, equipment inspection, respirator maintenance and repair, and program management, review, and enforcement.

A written respiratory protection program is required for any workplace where respirators are required to protect the health of the employees or whenever an employer requires employees to wear respirators. This is required by OSHA in both asbestos standards (29 CFR 1910.1001, 29 CFR 1926.1101) and the general respiratory protection standard (29 CFR 1910.134). The written respirator program establishes standard operating procedures concerning the use and maintenance of respiratory equipment. In addition to having such a written program, the employer must also be able to demonstrate that the program is enforced and updated as necessary.

The OSHA regulations detail what must be included in a written program. These items should include:

1. Procedures for selecting respirators for use in the workplace
2. Medical evaluations of employees required to wear respirators
3. Fit testing procedures for tight fitting respirators
4. Procedures for proper use of respirators in routine and reasonably foreseeable emergency situations
5. Procedures and schedules for cleaning, disinfecting, storing, inspecting, repairing, discarding, and otherwise maintaining respirators
6. Procedures to ensure adequate air quality, quantity, and flow of breathing air for atmosphere-supplying respirators
7. Training of employees in the respiratory hazards to which they are potentially exposed during routine and emergency situations
8. Training of employees in the proper use of respirators, including putting them on and removing them, any limitations to their use, and their maintenance
9. Procedures for regularly evaluating the effectiveness of the program

ESTABLISHING A POLICY

Every employer should prepare a clear concise policy regarding the use of respirators by their employees when performing asbestos abatement activities. This policy should serve as the guiding principal for the preparation, implementation, and enforcement of an effective respiratory protection program.

Designation of a Program Administrator

A program administrator must be designated by name. This person is responsible to administer or oversee the respiratory protection program. It is usually a good idea to also designate a person who is responsible for enforcement of the procedures at each job site. Procedures should also be outlined for enforcement of the program. Enforcement procedures and the development of the program as a whole should be done in conjunction and with input from the employees and/or their representatives.

SELECTION OF APPROPRIATE RESPIRATORY PROTECTION EQUIPMENT

The selection of an appropriate respirator generally involves three steps:

- Identifying the hazards;
- Evaluating the hazards;
- Providing proper respiratory protective equipment to suit the conditions and the individual.

Both the respirators selected and the respiratory program established must conform to Occupational Safety and Health Administration (OSHA) standards and guidelines published by respiratory manufacturers. The OSHA respirator standard (29 CFR 1910.134) requires that only approved respirators be used. They must be approved for protection specifically against asbestos fibers.

The National Institute for Occupational Safety and Health (NIOSH) is the official testing and approval agency for respirators and cartridges. If the entire respirator assembly including face piece, filters, hoses, and airlines pass the NIOSH criteria, then NIOSH issues an approval.

NIOSH currently rates negative pressure air-purifying respirator cartridges based on three efficiency levels. These levels are listed below:

Rating	Efficiency
95	95 % of particles
99	99 % of particles
100	99.97 % of particles at 0.3 Γm

Additionally, NIOSH rates air purifying particulate respirator cartridges in accordance with resistance to oil mists and solvents. These ratings are listed in the following table:

Rating	Oil Mists/ Solvents Resistance
N	Not resistant
R	Resistant, up to 1 shift
P	Oil proof, resistant for more than 1 shift

Asbestos protection requires the use of a 100 filter. The oil mists and solvents protection requirement will have to be determined based on evaluations at each job site. Asbestos projects may have the choice of using N-100, R-100, or P-100 filters. P-100 filters offer the most wide-spread protection.

Protection Factors

Respirators offer varying degrees of protection against asbestos fibers. The key to understanding the difference between types of respirators (air-purifying, powered air-purifying, air-supplied) is the amount of protection afforded the wearer. To compare these, one must understand the concept of a protection factor (PF).

The protection factor is defined as the concentration of a contaminant measured outside the mask divided by the concentration you would expect to find inside the mask. This simple formula is illustrated below.

$$\textbf{Protection Factor (PF)} = \frac{\textbf{Concentration Outside Mask}}{\textbf{Concentration Inside Mask}}$$

The actual level of protection depends greatly on the fit of the mask to the wearer's face. Accordingly, the protection offered by any one respirator will be different for each individual person. Further, the protection constantly changes depending upon the worker's activities and even shaving habits. When a worker laughs or coughs inside a respirator, the protection factor will decrease since the mask will not "fit" as well during laughing or coughing. Similarly, the worker who forgot to shave one morning will not receive as much protection that day since the mask will not fit as well to the face. The importance of properly fitting the mask should now be obvious.

It is virtually impossible to measure the concentration inside the mask (where the worker is breathing) for each worker, all the time, during the various activities he or she may be conducting. Accordingly, PFs based on extensive research have been developed for different categories of respirators. For a given class of respirator, the assigned PF will apply to that device when worn by a trained wearer on any given work day.

Using these PFs, it is easy to determine what type of respirator is appropriate to maintain the concentration of asbestos fibers inside the mask below a specified level. The assigned protection factors in Table 1 are only effective when the employer implements a continuing, effective respirator program as required by this section (29 CFR 1910.134), including training, fit testing, maintenance, and use requirements.

The following tables list the respirators OSHA approves for various types of activities with asbestos. The protection factor assigned by OSHA, OSHA's maximum airborne concentrations of asbestos or conditions of use, and the recommended maximum airborne concentrations for each respirator is provided.

PF = 10 if Qualitative Fit Testing performed:
PF = 50 if Quantitative Fit Testing performed
Recommended maximum use concentration based on NIOSH PF and achieving 0.01 f/cc inside mask.

Respirator Selection

All Activities except Removal of TSI and/or Surfacing Material Expected to Exceed 0.1 f/cc
(Includes OSHA Class II, III, IV asbestos work and other operations)

RESPIRATOR	OSHA	OSHA ALLOWABLE MAXIMUM USE CONCENTRATION	RECOMMENDED MAXIMUM USE CONCENTRATION
Half-Mask Air Purifying w/HEPA Filters	10	1.0 f/cc	0.1 f/cc
Full Face piece Air-Purifying w/HEPA Filters	50/10*	0.5 f/cc	0.5 f/cc
Powered Air-Purifying (PAPR) or Supplied Air, Continuous Flow Loose-Fitting Helmet or Hood with HEPA Filters	100 (NIOSH = 25)	10 f/cc	0.25 f/cc**
Powered Air-Purifying (PAPR), Full Face piece w/HEPA filters	1000	100 f/cc	10 f/cc
Supplied Air, Continuous Flow, Full Face piece + HEPA Filters	100	10 f/cc	1.0 f/cc
Full Face piece Supplied Air, Pressure Demand + HEPA Filters (Escape)	1,000	100 f/cc	10 f/cc
Full Face piece Supplied Air, Pressure Demand w/Auxiliary SCBA, Pressure Demand or Continuous Flow	>1,000	> 100 f/cc	> 10 f/cc

RESPIRATOR SELECTION

Removal of TSI and/or Surfacing Material with Expected Exposure over 0.1 f/cc (OSHA Class I asbestos work)

RESPIRATOR PF	OSHA PF	OSHA ALLOWABLE MAXIMUM USE CONCENTRATION	RECOMMENDED MAXIMUM USE CONCENTRATION
Powered Air-Purifying (PAPR), Full Face piece w/HEPA filters	1000	1.0 f/cc	1.0 f/cc
Full Face piece Supplied Air, Pressure Demand + HEPA Filters (Escape)	1000	1.0 f/cc	1.0 f/cc
Full Face piece Supplied Air, Pressure Demand w/Auxiliary SCBA, Pressure Demand or Continuous Flow	>1,000	>1.0 f/cc	>1.0 f/cc

Assigned Protection Factors (APF)

APF: the workplace level of respiratory protection that a respirator or class of respirators is expected to provide employees when the employer implements a continuing, effective respiratory protection program as specified by 29CFR 1910.134

- Must use the APF's listed to select a respirator that meets or exceeds the required level of protection
- When using a combination respirator (e.g., airline with an air-purifying filter), must ensure that the APF is appropriate to the mode of operation in which the respirator is being used

Maximum Use Concentrations (MUC)

MUC: the maximum atmospheric concentration of a hazardous substance from which an employee can be expected to be protected when wearing a respirator, and is determined by the assigned protection factor of the respirator or class of respirators and the exposure limit of the hazardous substance

- *MUC = APF* x *OSHA Exposure Limit*
- Must select a respirator that maintains exposure to the hazardous substance, when measured outside the respirator, at or below the MUC
- Must not apply MUCs to conditions that are IDLH; instead must use respirators listed for IDLH conditions per paragraph (d)(2)
- When the calculated MUC exceeds the IDLH level for a hazardous substance, or the performance limits of the cartridge or canister, then employers must set the maximum MUC at that lower limit
- Must select a respirator appropriate for the chemical state and physical form of the contaminant

PEL: the maximum allowable concentration level of a contaminant in the air to which an individual may be exposed for a time weighted average for an 8 hour day, 40 hour work week, set by OSHA

Respirators - Contaminant Maximum Concentration

Respirator	Maximum Use
APR Half Masks	10 times contaminant's PEL
APR Full Facepieces	50 times contaminant's PEL
PAPR Half Masks	50 times contaminant's PEL
PAPR Full Facepieces	1000 times contaminant's PEL
PAPR Loose Fitting Facepiece	25 times contaminant's PEL
PAPR Helmet / Hood	1000 times contaminant's PEL
SAR Half Mask - Cont. Flow	50 times contaminant's PEL
SAR Full Facepiece - Cont. Flow	1000 times contaminant's PEL
SAR Loose Fitting Facepiece	25 times contaminant's PEL
SAR Helmet / Hood	1000 times contaminant's PEL
SAR Half Mask PDAR	50 times contaminant's PEL
SAR Full Facepiece PDAR	1000 times contaminant's PEL
SCBA - Demand Mode	10000 times contaminant's PEL

Table 1. -- Assigned Protection Factors

Type of respirator[1], [2]	**Quarter mask**	**Half mask**	**Full facepiece**	**Helmet/ hood**	**Loose-fitting face-piece**
1. Air-Purifying Respirator	5	[3]10	50	....	....
2. Powered Air-Purifying Respirator (PAPR)	....	50	1,000	[4]25/1,000	25
3. Supplied-Air Respirator (SAR) or Airline Respirator					
• Demand mode	....	10	50	....	....
• Continuous flow mode	....	50	1,000	[4]25/1,000	25
• Pressure-demand or other positive-pressure mode	....	50	1,000	....	....
4. Self-Contained Breathing Apparatus (SCBA)					
• Demand mode	....	10	50	50	....
• Pressure-demand or other positive-pressure mode (e.g., open/closed circuit)	....	....	10,000	10,000	....

[1]Employers may select respirators assigned for use in higher workplace concentrations of a hazardous substance for use at lower concentrations of that substance, or when required respirator use is independent of concentration.

[2]The assigned protection factors in Table 1 are only effective when the employer implements a continuing, effective respirator program as required by this chapter (29 CFR 1910.134), including training, fit testing, maintenance, and use requirements.

[3]This APF category includes filtering face pieces, and half masks with elastomeric face pieces.

[4]The employer must have evidence provided by the respirator manufacturer that testing of these respirators demonstrates performance at a level of protection of 1,000 or greater to receive an APF of 1,000.

This level of performance can best be demonstrated by performing a WPF or SWPF study or equivalent testing. Absent such testing, all other PAPRs and SARs with helmets/hoods are to be treated as loose-fitting face piece respirators, and receive an APF of 25.[5]These APFs do not apply to respirators used solely for escape. For escape respirators used in association with specific substances covered by 29 CFR 1910 subpart Z, employers must refer to the appropriate substance-specific standards in that subpart. Escape respirators for other IDLH atmospheres are specified by 29 CFR 1910.134 (d)(2)(ii).

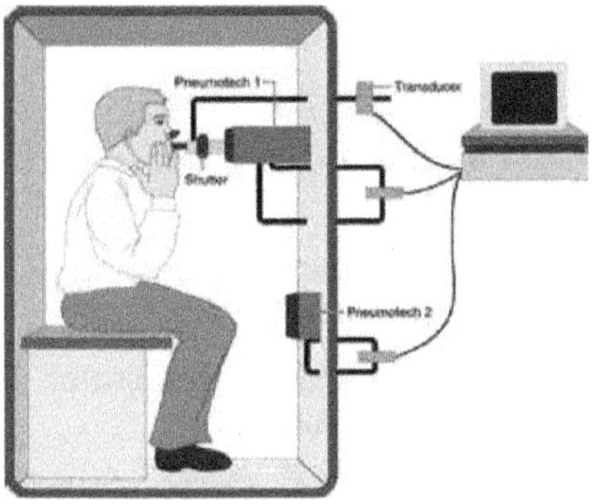

MEDICAL APPROVAL

Only those individuals who are medically capable to wear respiratory protective equipment shall be issued a respirator. Each employee using a respirator must be evaluated before the respirator is used on the job, and their medical status re-evaluated annually thereafter. Medical tests must be supervised by a physician often include: pulmonary function tests, a chest x-ray (if a physician deems it necessary), electrocardiogram, and any other tests needed for proper evaluation by a physician. A medical history in the form of a questionnaire is collected as well for each individual. Other factors to be considered by a physician may include: emphysema, asthma, chronic bronchitis, heart disease, anemia, hemophilia, poor eyesight, poor hearing, hernia, lack of finger or hand usage, epileptic seizures, and other factors that might inhibit the ability of an employee to wear respiratory equipment.

EMPLOYER TRAINING PROGRAM

Each worker designated to wear a respirator must receive adequate training. The training sessions (initial and periodic training – at least annually) should be conducted by a qualified individual to ensure that employees understand the limitations, use, and maintenance of respiratory equipment. Training should take into account the educational level of the wearers, possible language barriers, and should be made as hands-on as possible.

RESPIRATOR FITTING

One of the most important elements of an effective program is proper respirator fit. The OSHA Respiratory Protection Standard (29 CFR 1910.134) requires that the fit of all tight-fitting respirators; either required to be worn by an OSHA standard, or required by the employer, be tested. Fit testing is required prior to initial use and at least annually thereafter. Procedures for fit-testing are contained in Appendix A of the OSHA Respiratory Protection Standard and must be addressed in the written respiratory protection program.

Each make, model, and size of tight-fitting respirator must be fit-tested. Fit- testing must also be performed when there is a noticeable change to the employee's physical condition. A respirator will not protect the wearer unless the air actually passes through the filter or canister, or unless all of the air comes from the supply system. If the face-to-face piece is not tight or the air hose connections are loose, the wearer may actually be breathing contaminated air from the work environment.

A person may have to try several different respirators before finding one that fits properly. Men may have to shave beards and bushy sideburns in order to wear a tight-fitting respirator, because the face piece does not seal over facial hair. Similarly, gum and tobacco chewing cannot be permitted since excess facial movement can break the face seal. Those who wear prescription glasses must wear a respirator face piece that will accommodate vision correction.

The OSHA also requires that the fit of respirators be checked each time the respirator is worn according to the manufacturer's instructions. This procedure is called a user seal check.

Fit testing falls into two major categories: qualitative (pass or fail test) and quantitative (measurement of contaminant levels inside the mask). Only those tests applicable to asbestos work are discussed below.

User Seal Checks

Once the respirator has been selected and no visual leaks are evident a negative pressure seal check and positive pressure seal check are performed by the wearer. These simple procedures are described below.

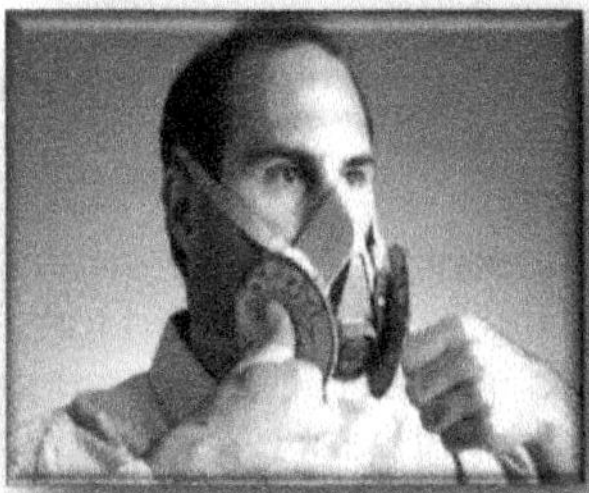

Negative Pressure Seal Check
For this fit check, the wearer:

- closes off the filter or cartridge inlets by covering them with the palms of the hands, or by blocking the breathing hose so that air cannot pass through;
- inhales so that the face piece collapses slightly; and holds his/her breath for about 10 seconds.

If the face piece remains slightly collapsed and no inward leakage of air is detected, the respirator passes the check. This check can only be used on respirators with tight fitting face pieces. Its potential drawback is that hand pressure can modify the face piece seal and cause false results.

Positive Pressure Seal Check

This fit check is similar in principle to the negative pressure seal check. The wearer:

- closes off the exhalation valve of the respirator; and
- gently exhales into the face piece for about 10 seconds.

The respirator fit is considered passing if positive pressure can be built up inside the face piece without evidence of outward air leakage around the face piece. If the selected respirator fails to pass the user seal checks, the wearer should readjust the straps and try again. If it still fails, the condition of the respirator should be checked to ensure that no damage has occurred. If the condition of the respirator is acceptable and a seal cannot be attained, another size or another brand should be donned and these checks repeated. Once the wearer has successfully passed the negative and positive pressure fit checks, the actual fit test may be conducted.

Qualitative Fit Testing

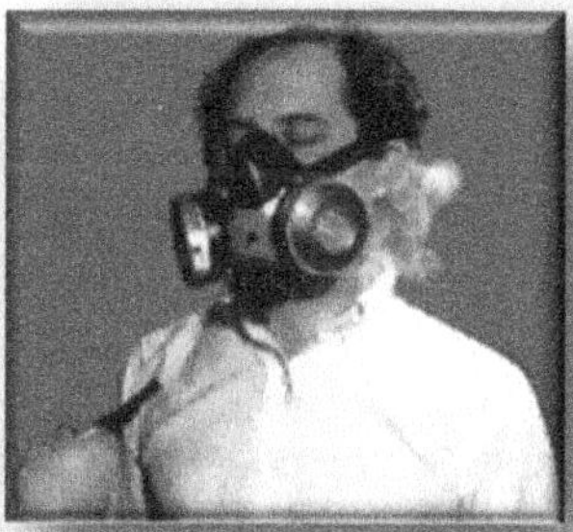

Qualitative fit testing involves the use of contaminants and test exercises to determine whether a respirator's fit is sufficient. If the wearer detects the contaminant during the course of the fit testing exercises, the respirator fit is deemed not sufficient and not approved. If the contaminants are not detected during the fit test exercises, the respirator fit is deemed satisfactory until the next fit test. Four (4) test solutions are currently permitted by OSHA as contained in Appendix A of the Respiratory Protection Standard. The following table lists the contaminants with advantages and disadvantages for each.

Quantitative Fit Testing

The qualitative fit test procedures involve two stages of testing. The first stage is a screening which involves exposing the test subject to low concentrations of the selected contaminant to ensure that the subject is sensitive to the solution. The second stage is the fit test which includes exercises testing the respirator's fit. These exercises are conducted under a hood or test enclosure for the banana oil, saccharin, and Bitrex. The test exercises are not permitted to be conducted under a hood during the irritant smoke test.

During fit testing, the respirator straps must be properly adjusted, in accordance with the manufacturer's directions, and should be as comfortable as possible. Over-tightening the straps will sometimes reduce face piece leakage, but the wearer may be unable to tolerate wearing the respirator for any length of time.

Solution	Advantages	Disadvantages
Isoamyl Acetate (Banana Oil)	Not harmful to human health	Does not cause involuntary reactions Requires use of organic vapor filter
Saccharine	Not harmful to human health	Does not cause involuntary reactions
Bitrex	Not harmful to human health; Causes involuntary	None evident
Irritant Smoke (Stannic Chloride)	Causes involuntary reaction	Can be harmful to human health (carcinogen)

The face piece should not press into the face and shut off blood circulation or cause any major discomfort. At the time of respirator selection, a visual inspection of the fit should always be made by a second person.

For the test, the respirator wearer enters a test enclosure (except for irritant smoke) and the test solution is sprayed or squeezed near the respirator wearer's head. The person then performs eight exercises in order, each for one minute, that simulate activities normally encountered in a work situation:

- normal breathing
- deep breathing
- turning head from side to side
- nodding head up and down
- talking
- grimace
- bending over or jogging
- normal breathing

Quantitative Fit Testing

Quantitative fit testing requires specialized equipment to measure differences in airborne particulate concentrations and a trained tester. Equipment and methods approved by OSHA, under the respiratory protection standard, include; condensation nuclei counter, Portacount™, and controlled negative pressure (CNP). The airborne concentration of particulate is measured both outside the respirator and inside the respirator (through a probe) while the worker performs several exercises.

Some tests require the introduction of contaminants, while others measure ambient air particulate. The specific degree of protection – fit factor – is determined for the wearer with the specific respirator worn. Fit factor is different from protection factor in that the fit factor is unique to the specific respirator worn by a specific individual at the time of the test. The minimum fit factor required for half-mask respirators is 100 and the minimum fit factor required for full face respirators is 500. These levels must be achieved or surpassed during the test exercise.

Fit Test Method Selection

Determining which fit test method to use is based on test quality and costs. The quantitative fit test is viewed as a better method than the qualitative fit test. The accuracy of the probes used to measure levels inside and outside the respirator is superior to relying on the detection of a test solution by the respirator wearer. However, the cost of quantitative fit test equipment is quite expensive [Check with your instructor on current prices]. This difference often causes employers to consider the quantitative fit tests as cost prohibitive, resulting in use of the qualitative fit test procedure. Additionally, OSHA has fit test protocol requirements based on the type of respirator. Quantitative fit testing is always approved, but for some respirator types, qualitative fit testing limits the protection factor a respirator is assigned or may not be permitted by OSHA. A list of respirators and the approval for quantitative fit testing (QNFT) or qualitative fit testing (QLFT) is provided in the following table.

Acceptable Fit Test Methods

Respirator	QNFT	QLFT
Half-face, Negative pressure, APR PF = 10	Yes	Yes
Full-face, Negative pressure, APR PF = 50	Yes	No
PAPR PF = 1000	Yes	Yes (Neg. Pressure
Supplied air, Demand mode PF = 100	Yes	Yes (Neg. Pressure
Supplied air, Pressure demand mode PF = 1000	Yes	Yes (Neg. Pressure
Loose fitting respirators (hood, helmet)	Fit testing not required	

CLEANING AND DISINFECTION OF RESPIRATORS

Whenever possible, a respirator should be reserved for the exclusive use of a single individual. After each use, the respirator should be cleaned and disinfected as follows:

- Wash the mask body in warm water using a brush with a detergent or a detergent/disinfectant combination.
- Rinse it in clean water, or rinse it once with a disinfectant and once with clean water. The clean-water rinse is particularly important because traces of detergent or disinfectant left on the mask can cause skin irritation and/or damage to respirator components.
- Air dry the respirator on a rack or hang it in a position that does not distort the shape of the elastomeric face piece.

ROUTINE INSPECTION OF RESPIRATORS

Respirator inspection is an important, routine task that should be performed before and after each use. The following items must be checked, at a minimum:

Air-Purifying Respirators (half-mask and full face piece)

The face piece should be checked for:

- Excessive dirt;
- Cracks, tears, or holes;
- Distortion from improper storage;
- Cracked, scratched, or loose fitting lens; and
- Broken or missing mounting clips.

Head straps should be checked for:

- Breaks or tears;
- Loss of elasticity;
- Broken or malfunctioning buckles or attachments; and
- Excessively worn serrations on the head harness which might allow the face piece to slip.

The inhalation and exhalation valves should be checked for:

- Detergent residue, dust particles or dirt on the valve seat;
- Cracks, tears, or distortion in the valve material or valve seat; and
- Missing or defective valve cover.

Filter elements should be checked for:

- Proper filter for the hazard;
- Approval designation (TC_ID #_);
- Missing or worn gaskets;
- Worn threads; and
- Cracks or dents in filter housing.

Powered Air Purifying Respirators

Check face piece, head straps, valve, and breathing tube, as for regular air purifying respirators.

Hood or helmet, if applicable -- check for:

- Headgear suspension (adjust properly for wearer); and
- Cracks or breaks in face shield (replace face shield).

Supplied Air Respirators

Face piece, head strap, and valves should be checked as specified above. In addition, the following checks should be performed:

Breathing tube should be checked for:

- Cracks;
- Missing or loose hose clamps; and
- Broken or missing connectors.

Hood, helmet, or suit should be checked for:

- Headgear suspension;
- Cracks or breaks in face shield; and
- Rips and torn seams.

Air supply systems should be checked for:

- Breaks or kinks in air supply hoses and end fitting
- Tightness of connections;
- Proper setting of regulators and valves (consult manufacturers recommendations); and
- Correct operation of air purifying elements and carbon monoxide or high-temperature alarms.

Self-contained Breathing Apparatus (SCBA)

Consult manufacturer's literature.

MAINTENANCE AND REPAIR

At some point any respirator will need replacement parts or some other repair. Only trained, qualified persons can legally repair respirators. Respirator parts from different manufacturers are **NOT** interchangeable. NIOSH approval is invalidated if parts are substituted, or non-approved parts are used.

RESPIRATOR STORAGE

Proper storage is very important. Respirators must be protected from dust, sunlight, heat, extreme cold, excessive moisture, and damaging or contaminating chemicals. When not in use, the respirator should be placed in a closed, re- sealable plastic bag, and stored in a clean, convenient, sanitary location.

SURVEILLANCE OF WORKING CONDITIONS

The employer must provide adequate surveillance of the employees' working conditions to be certain the respirator selected provides adequate protection. In the case of asbestos abatement, the employer must also determine if workers might encounter other hazardous airborne contaminants for which the respirator chosen is not adequate. Air monitoring to estimate the asbestos exposure provides the needed information to determine if the selected respirator provides sufficient protection to the wearers.

RESPIRATOR PROGRAM EVALUATION AND RECORDKEEPING

The employer is also responsible for evaluating the respirator program at least once annually and make program adjustments, as appropriate, to reflect air sampling or other evaluation results. The employer should also review compliance with all aspects of the program outlined in this chapter (respirator selection, purchase of approved equipment, medical evaluation of employees, fit testing, issuance of equipment and associated maintenance, storage, repair and inspection, appropriate surveillance of work area conditions).

Attention should be given to proper recordkeeping. Records that should be kept include:

- names of employees who have been trained in respirator use;
- documentation of the care and maintenance of respirators;
- medical reports of each respirator user;
- test results showing possible airborne concentrations of asbestos fibers during work; and
- instances of any problems specific to the use of respirator equipment encountered during work.

Protective Clothing

The primary reason people wear protective clothing during asbestos abatement is to keep gross amounts of asbestos-containing debris off the body, hair, etc. The use of protective clothing and showers will minimize the chance of bringing asbestos out of the work area and into the home. Protective clothing will also minimize the chance of rashes and discomfort caused by the material being removed. In addition to the asbestos, frequently the material being removed contains mineral wool, fiberglass, and binders such as cement. Each of these may be irritating to the skin. Continued direct contact with asbestos has also been shown to cause "asbestos warts."These warts often take months to heal and occur more frequently if asbestos is trapped beneath a watchband, or in other ways kept in close contact with the skin.

Protective clothing for asbestos abatement projects usually consists of disposable coveralls, foot covering and head covering. The foot and head covering should be attached to the coveralls. This eliminates the need to tape openings between garments, etc. Tight fitting bathing suits may sometimes be worn beneath the coveralls. Nylon suits work well and can be cleaned easily during showering. Gloves should be worn when inside the work area. Any article that cannot be decontaminated should remain on the dirty equipment room side of the enclosure.

Protective clothing does NOT include street clothes, T-shirts, blue jeans, sweat bands, kneepads and socks. If any of these items are used inside the work area, they should remain there until the job is completed and be disposed of as asbestos-contaminated waste. Similarly, jewelry such as rings and ID bracelets should not be worn in the work area.

Other protective clothing and items such as hard hats and safety shoes or boots should remain in the work area for the duration of the project. Upon project completion, these items can be cleaned, placed in a plastic bag, labeled as containing asbestos, and taken to the next project. If safety shoes or boots are not worn, it is wise to have workers wear rubber soled, slip-on deck shoes. These remain in the work area and are disposed of at the end of the project as asbestos-containing waste. These deck shoes are usually of canvas construction and are inexpensive (about $10.00 per pair). It is a good idea to have each worker mark their name on shoes and hard hat with permanent ink. Employers should provide and ensure the use of head and foot protection in accordance with OSHA standards 29 CFR 1910.135 and 1910.136 respectively.

To summarize, below is a list of items normally worn by asbestos abatement workers:

- Disposable coveralls
- Disposable foot covering
- Disposable head covering
- Tightly-fitting nylon swimsuit or disposable undergarments
- Waterproof safety shoes or boots (as required)
- Hard hat (as required)
- Gloves (cotton with Kevlar™ or other protective material on the palm)
- Eye protection (not needed if full face piece respirators are used)

The disposable coveralls, foot, and head coverings are available from many sources and several materials. Check with your instructor on current prices. It is important to realize that many "bargain" prices may not be a bargain at all. The less expensive coveralls often use less material. Some coveralls available in one size only may be too small for many workers. Be sure to check the construction of the coveralls as well. Double stitching on seams costs more but will last longer.

A common problem on asbestos abatement projects is a failure by contractors to purchase enough coveralls for the project. Each worker must use a new coverall (and foot and head covering if not attached) each time he/she enters the work area. Assuming two breaks and a lunch period, four coveralls will be needed each day by each worker.

Additional coveralls are usually needed for authorized visitors (architect, industrial hygienist, etc.) and to replace some that are torn to the point of being unusable. As a rule of thumb, the contractor may estimate the number of suits needed on a project by the following formula:

5 x number of workers x project duration in days = number of coveralls needed

For example, a project lasting 48 days using a crew of eight workers and one job foreman will need the following number of coveralls (estimated):

5 x 9 workers x 48 days = 2,160 coveralls

Accordingly, the contractor should order 90 cases (24 per case) of coveralls for the project. A prudent contractor would purchase 100 cases to allow for sufficient surplus. When purchasing coveralls, large and extra large sizes should be purchased. These can always be made to fit smaller employees.

PUTTING ON PROTECTIVE CLOTHING

Protective clothing is put on in the clean room of the decontamination unit before entering the work area using the following sequence:

1. Remove all street clothes (including undergarments and jewelry) and store them in a clean, convenient location such as a bin or locker. Note: It is usually wise to have a lockbox or other means to protect valuables and to discourage employees from bringing wallets, rings, keys, etc., into the work area.
2. Put on the nylon swimsuit or disposable undergarments, d.
3. Put on the disposable overalls.
4. If separate disposable foot coverings are used, these are put on.
5. Tape the ankles to take up slack in the suits and reduce the chance of tripping. (Tape pants over foot coverings, if separate.)
6. Inspect the respirator equipment, put it on, and conduct fit checks.
7. Put on the hood or head covering over the respirator head straps.
8. Pass through the airlocks and shower room into contaminated equipment room.
9. Put on safety shoes or boots, as required.
10. Put on gloves (cotton gloves with Kevlar are usually worn although leather gloves should be used for handling metal lath). and tape the sleeves over the gloves using duct tape.
11. Put on other protective equipment, such as hard hats and safety glasses (if a half-face respirator is used).

One person should remain outside the work area at all times. It should be his/her responsibility to ensure that each person entering the work area has the proper protective clothing and to log them in and out.

Once inside the work area, no workers, or anyone else, should be permitted to leave without going through the decontamination sequence unless it is an extreme emergency. A common problem is employees **"stepping out"** for a cigarette or supervisors **"stepping in"** the work area to deliver a message or a piece of equipment. These activities defeat the purpose of the protective equipment and the decontamination sequence.

TAKING PROTECTIVE CLOTHING OFF

Whenever an employee or other person leaves a work area for any reason, he/she must go through the decontamination sequence. This sequence should include the following steps:

1. Clean reusable protective equipment such as boots/shoes, safety glasses, hard hats, etc.
2. Remove all protective garments and equipment (except respirators) in an area immediately outside the shower on the contaminated side. An area should be designated for this purpose and kept as free as practicable of asbestos-contaminated material. All clothing should be placed in plastic bags inside a drum and labeled as asbestos-containing waste.
3. Proceed to the shower while still wearing respirator. The worker should wash his/her entire body, including head, hair, and face (except what is under the respirator straps) while still wearing, and breathing through, the tightly fitting respirator. This should be done without soaking the cartridges. After the entire body has been decontaminated, remove the respirator, soak the cartridges, and finish showering. The cartridges may then be discarded in a plastic bag located at the shower.
4. Proceed to the clean room, dry off, dress in street clothes, and disinfect, clean, and inspect respirator. If air supply is not being used, new cartridges should be placed in the respirator.

OTHER PERSONAL PROTECTIVE EQUIPMENT

Additional protective equipment may be necessary depending on the specific project. The most common other protective equipment will include eye protection such as goggles or safety glasses with side shields. Hard hats, safety shoes, and hearing protection may also be necessary on certain projects.

RESPIRATOR PROGRAM CHECKLIST

In general, the respirator program should be evaluated at least annually with program adjustments, as appropriate, made to reflect the evaluation results. Program function can be separated into administration and operation.

A. Program Administration

__✓__ Is there a written policy which acknowledges employer responsibility for providing a safe and healthful workplace, and assigns program responsibility, accountability and authority?

__✓__ Is program responsibility vested in one individual who is knowledgeable and who can coordinate all aspects of the program at the jobsite?

__✓__ Can feasible engineering controls or work practices eliminate need for respirators?

__✓__ Are there written procedures/statements covering the various aspects of the respirator program, including:

- ✓designation of an administrator
- ✓respirator selection
- ✓purchase of approved equipment
- ✓medical aspects of respirator usage
- ✓issuance of equipment
- ✓fitting
- ✓training
- ✓maintenance, storage and repair
- ✓inspection
- ✓use under special conditions and
- ✓work area under surveillance?

B. <u>Program Operation</u>

(1) **Respiratory protective equipment selection and assignment**

______ ✓ Are work area conditions and employee exposures properly surveyed?

______ ✓ Are respirators selected on the basis of

✓ hazards to which the employee is exposed?

______ ✓ Are selections made by individuals knowledgeable of proper selection procedures?

✓ Are only approved respirators purchased and used; do they provide adequate protection for the specific hazard and concentration of the contaminant?

✓ Has a medical evaluation of the prospective user been made to determine physical and psychological ability to wear the selected respiratory protective equipment?

______ ✓ Where practical, have respirators been issued to the wearers for their exclusive use, and are records covering issuance kept?

(2) **Respirator fitting**

______ ✓ Are the users given the opportunity to try on several respirators to determine whether the respirator they will subsequently be wearing is the best fitting one?

______ ✓ Is the fit tested at appropriate intervals?

______ ✓ Are those users who require corrective lenses properly fitted?

______ ✓ Are users prohibited from wearing contact lenses when using respirators?

______ ✓ Is the face piece-to-face seal tested in a test atmosphere?

______ ✓ Are workers prohibited from entering contaminated work areas when they have facial hair or other

✓ characteristics which prohibit the use of tight-fitting face pieces?

______ (3) **Respirator use**

______ ✓ Are respirators being worn correctly (i.e., head covering over respirator straps)?

______ ✓ Are workers keeping respirators on all the time when necessary?

(4) **Maintenance of respiratory protective equipment**

(a) Cleaning and Disinfecting

______ ✓ Are respirators cleaned and disinfected after each use?

______ ✓ Are proper methods of cleaning and disinfecting utilized?

(b) Storage

______ ✓ Are respirators stored in a manner so as to protect them from dust, sunlight, heat, excessive cold or moisture, or damaging chemicals?

✓ Are respirators stored properly in a storage facility so as to prevent them from deforming?

✓ Is storage in lockers and tool boxes permitted only if the respirator is in a carrying case or carton?

(c) Inspection

______ ✓ Are respirators inspected before and after each use and during cleaning?

______ ✓ Are qualified individuals/users instructed in inspection techniques?

______ ✓ Is respiratory protective equipment designated as "emergency use" inspected at least monthly (in addition to after each use)?

______ ✓ Is a record kept of the inspection of "emergency use" respiratory protective equipment?

(d) Repair

______ ✓ Are replacement parts used in repair those of the manufacturer of the respirator?

(5) **Special use conditions**

_____ ✓ Is a procedure developed for respiratory protective equipment usage in atmospheres immediately dangerous to life or health?

_____ ✓ Is a procedure developed for equipment usage for entry into confined spaces?

(6) **Training**

_____ ✓ Are users trained in proper respirator use, cleaning and inspection?

_____ ✓ Are users trained in the selection of respirators?

_____ ✓ Are users evaluated, using competency-based evaluation, before and after training

Objectives: To become familiar with the equipment, procedures, use, limitations, and maintenance of a supplied-air respiratory protection system.

Learning Tasks: Information in this chapter should enable participants to:

- Understand the reasons for using air-supplied respirators.
- Recognize that breathing air must be processed and not simply pumped.
- Understand the mechanism by which air is compressed and purified.
- Become familiar with the need for air storage banks and proper air delivery to the worker.
- Gain an understanding of the many regulations and recommended practices for providing breathing air.

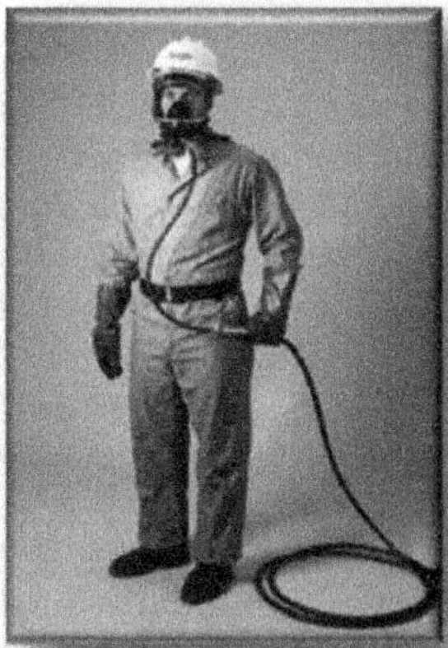

ESTABLISHING A TYPE C SUPPLIED-AIR SYSTEM

Good practices generally require the use of type C supplied-air respirators on any asbestos removal project. If the assumption is made that 0.01 fibers per cubic centimeter (f/cc) will be the clearance level at the conclusion of a removal project, the workers performing the abatement work should not be exposed, inside the respirator, to concentrations of asbestos fibers any greater than this. The selection of appropriate respiratory protection now becomes greatly simplified. Using established protection factors (a ratio of the concentration outside the mask to that expected inside the mask); maximum anticipated fiber levels outside the respirator may be used to select the proper respiratory protection to reduce the concentration inside the mask to 0.01 f/cc or below.

As a rule of thumb, air-purifying respirators are usually adequate protection for the worker during work area preparation and final wipe-down following gross clean-up. Good practices require the use of type C supplied-air respirators once gross removal begins through gross clean up of the work area.

A type C supplied-air system normally consists of a compressor, air delivery lines, air cleaning apparatus, a reserve air supply, and NIOSH-approved masks. Instead of a compressor, the source of compressed air may also be a bank of high pressure air cylinders, in which case, an air cleaning apparatus is not necessary. At a minimum, a type C system should provide the following:

- Continuous and sufficient supply of Grade D air;
- NIOSH-approved respirators and supply hoses;
- Adequate reserve for escape time;
- Breathing air temperature control;
- Continuous monitor and alarm for carbon monoxide (CO).

GRADE D AIR

Grade D air is the minimum quality for routine use in supplied-air (or self-contained) breathing equipment, as used in firefighting, general industry, and asbestos abatement projects where supplied-air respirators are in use. There are other grades of air purity including Grade E (minimum requirements for sports diving to 125 feet) and Grade H. Each of these grades (E and H) is more stringent than Grade D specifications. The Grade D air specifications were established by the Compressed Gas Association, Inc. of New York and incorporated into the OSHA Respirator Standard (29 CFR 1910.134) by reference. The specifications themselves are contained in the Compressed Gas Association (CGA) Pamphlet G-7 entitled, "Compressed Air for Human Respiration." These specifications are discussed briefly below. Asbestos abatement contractors performing work in Canada should be aware that breathing air must meet considerably more stringent standards as described in the Compressed Breathing Air Standard Z180.

GRADE D BREATHING AIR REQUIREMENTS

Oxygen	19.5 - 23.5%
Carbon Monoxide	(CO 10 parts per million, maximum
	(10 ppm under CGA[1])
Carbon Dioxide (CO_2)	1000 parts per million, maximum
Condensed Hydrocarbons	5 milligrams per cubic meter, maximum
Objectionable Odors	None
Water Vapor	The CGA standard does not specifically establish a limit for moisture; however, a limit of 66 parts per million is necessary to assure proper function of CO scrubbing devices.

Normal air contains 20.9% oxygen. The oxygen content in breathing air should always fall between 19.5% and 23.5%. Normally, the oxygen content is only a consideration when purchasing bottled air which has been manufactured. Since the oxygen content of ambient air remains quite constant, and compressing the air does not alter the oxygen content, there is little concern that the asbestos abatement worker will be short of oxygen.

Perhaps the greatest concern when dealing with type C supplied-air systems is the generation or presence of carbon monoxide. This contaminant may be introduced into the breathing air through compressor malfunction or, more commonly; it may be drawn into the compressor intake. Carbon monoxide may be produced by the compressor if it overheats. The overheating causes the lubricating oil to break down with carbon monoxide being released. For this reason, high temperature alarms are often installed on compressors. OSHA requires that oil-lubricated temperature alarm is used; the air from the compressor shall be frequently tested for carbon monoxide. The term "frequently" is not defined by OSHA, but a General Services Administration Specification for Breathing Air (BB-A-1034a) requires manual CO testing after every 4 hours of use.

One alternative is to use an oil-free compressor to eliminate the chance of oil breakdown if the compressor overheats. However, oil-free compressors usually require more frequent servicing, and the synthetic materials used may release gaseous contaminants if the compressor overheats. (Note: The Canadian Standards Association has established a limit of 2 ppm in compressed breathing air for each of the following: trichlorotrifluoroethane, dichlorodifluoromethane, and chlorodifluoromethane.)

To avoid drawing carbon monoxide into the compressor directly, an extension intake flexible duct should be used to place the air intake at a remote location. The location chosen should be away from any combustion sources (i.e., vehicle exhausts, smokestacks, etc.). Frequently, the best location is 15 or 20 feet up in a tree, since it would be unlikely that a truck or car, lawnmower (or other carbon monoxide producing vehicle could affect the supply air) coarse filter (screen) over the air inlet to keep leaves, bugs, etc., from being drawn into the compressor.

Contractors/supervisors should be aware of any other potential sources of toxic gases near the air intake. This would be especially important in industrial settings where gases are commonplace. Contaminants not listed in the specifications for Grade D air should not exceed one-tenth of the Threshold Limit Values (TLVs) for Chemical Substances in the work Environment, adopted by the American Conference of Governmental Industrial Hygienists (ACGIH).

AIR PROCESSING

A properly established type C supplied-air system does not simply pump air to worker. The air must also be processed. Along with air that enters the compressor, heat and water vapor are also processed. As the air is compressed, the temperature rises. When the compressed air drops back to normal ambient pressure, the temperature likewise returns to normal. Therefore, heat should be removed from the compressed air so that the air that reaches the worker inside the respirator is cool and comfortable.

Water vapor, when compressed, forms water droplets or condensation. If this water is not removed, it can build up in the air lines to the workers to the point where a solid "plug" of water is formed. This plug of water will quickly be forced into the respirator of the workers. It is quite likely that the workers will immediately discard the masks, or be startled by the sudden flood of water, potentially causing an accident (fall from a scaffold or ladder). Accordingly, the air processing equipment must be capable of removing moisture from the supply air.

There are four basic steps in establishing a type C supplied-air system:

- compression
- purification
- storage
- distribution

Compression

Compression of air is necessary to store the air in a reduced space until needed. There are many different types of compressors available to perform this task. Some are oil-free using non-lubricated Teflon piston rings. More typical is the reciprocating multistage oil-lubricated compressor. Screw-type compressors and diaphragm compressors are also available. Compressors may be gasoline, diesel, or electric powered. In a reciprocating compressor, air is compressed in steps or stages. At each stage (or immediately following) there should be a condensate trap to allow water to be expelled. The compressor should have a high air temperature shutdown switch, an automatic start-stop pressure switch, and a low oil level shutdown switch. The choice of lubricating oil will depend on the individual make and model of the compressor. Usually, however, mineral oil or a high grade synthetic oil is used. The compressor should also have appropriate pressure gauges and safety valves.

When selecting a compressor (or renting), be sure the manufacturer is aware of its intended use. Some compressors deliberately add oil to the air stream for lubricating air-driven machinery. Further, be sure to follow the manufacturer's operating instructions and service schedule.

The compressor should be located in a covered area with good ventilation. An area under a shade tree with a cool breeze is ideal, but the compressor should never be placed in an area with below freezing temperatures. The air intake should be located at least 8 feet above the ground and a coarse filter used as previously described. Gasoline should always be stored out of direct sunlight in a cool area, not accessible to visitors, etc.

Purification

The second step in air processing is purification. While the compressor may remove some water through the condensate traps, additional moisture, odors, oil, hydrocarbons, heat, and carbon monoxide must be removed. Equipment for the purification process consists of a series of filters, adsorption columns, and sieves. Check with your instructor as to the prices of both low pressure units, and high pressure units.

Purification equipment normally has an after cooler which is air or water-cooled to remove heat, thereby cooling the air. Following the after cooler, the air stream passes through one or more coalescing elements. A coalescing element causes aerosol droplets to join together (or coalesce) and form larger drops that impinge on the filtering surface and are too heavy to be re-entrained in the air stream. The coalescing element removes droplets of water and oil, as well as solid particles larger than about 10 micrometers in diameter through mechanical filtration.

Two adsorption canisters are usually located next in line to the coalescing element(s). The first adsorption canister consists of a column packed with a molecular sieve to remove water vapor. This canister also removes gaseous hydrocarbons, nitrogen oxides, sulfur compounds, and other odors. Following this canister is a column containing activated charcoal, which removes additional unpleasant odors and oil vapor. Each of the sorbent materials (molecular sieve and activated charcoal) will need periodic replacement according to the manufacturer's specifications.

The next step in the purification process is eliminating carbon monoxide through oxidation. In the presence of a catalyst, typically hopcalite, carbon monoxide will combine with oxygen to form carbon dioxide, a much less harmful gas. It should be noted that water destroys the catalyst. Therefore, the removal of water and water vapor before this stage of purification is extremely important.

Following the carbon monoxide catalyst, there is usually a mechanical filter to remove any particles larger than 0.5 micrometers in diameter. After the filter, there is usually a carbon monoxide monitor and alarm. This instrument measures the concentration of carbon monoxide in the supply air stream. The calibration procedure and frequency specified by the manufacturer should be followed (for example, the MSA CO monitor must be calibrated monthly). The carbon monoxide monitor should be equipped with a visual and audible alarm to alert the operator of a high carbon monoxide level in the supply air. Also, the filtration unit should have an air-powered horn to alert the operator of electrical power failure. The unit will continue to function; however, the carbon monoxide monitor and the after cooling fan will not operate without electricity.

Storage

Air exiting the purification unit should meet the Grade D requirements already discussed. If a high pressure supplied-air system (greater than about 200 psi) is used, the purified air may go directly into a high pressure storage tank or reservoir. A high pressure air line (usually a stainless steel, 1/4 to 1/2 inch inside diameter line) is run into the work area to a manifold and regulator. At this point, the regulator reduces the pressure to less than 125 psi for service to the workers through low pressure air lines. If a low pressure system (less than 200 psi) is used, the purified air will usually be fed directly to a manifold in the work area via low pressure (high volume) air line. At the manifold, several air lines will be run to the individual workers. To store air in the event of compressor shutdown, a check valve is installed between the filtration unit and the work area manifold. If the compressor shuts off, the check valve should open to provide air from a series of bottles (high pressure) containing grade D breathing air. The volume of air to provide adequate escape time will depend on the number of employees and time needed to evacuate a work area in an orderly manner. A reservoir of air for this purpose is required by OSHA (29 CFR 1910.134).

Delivery

Once the air has been compressed, purified, and an adequate reserve available for emergencies has been established, it is ready for delivery to the asbestos abatement workers. Usually, large air lines from outside the work area are used to feed manifolds in the work area. Each manifold can usually accommodate 2-6 air lines. In this way, short air lines (about 50 feet) can be used, reducing the amount to be pulled about by the worker or tripped over by others. Each airline connects to a belt- or face piece-mounted regulator which controls flow into the mask.

The amount of air needed will depend on many factors including the type of respirator, number of workers, and auxiliary equipment. Tight-fitting supplied-air masks must be supplied with a minimum of 4 cubic feet of air per minute (CFM). Loose-fitting supplied-air respirators must be provided with 6 CFM at a minimum. For each of these types of masks, the maximum recommended flow rate is 15 CFM. It should be noted that the use of vortex air cooling devices will require additional air flow according to the specifications of the vortex unit chosen.

Any air-line respirator chosen must be approved by NIOSH (National Institute for Occupational Safety and Health.

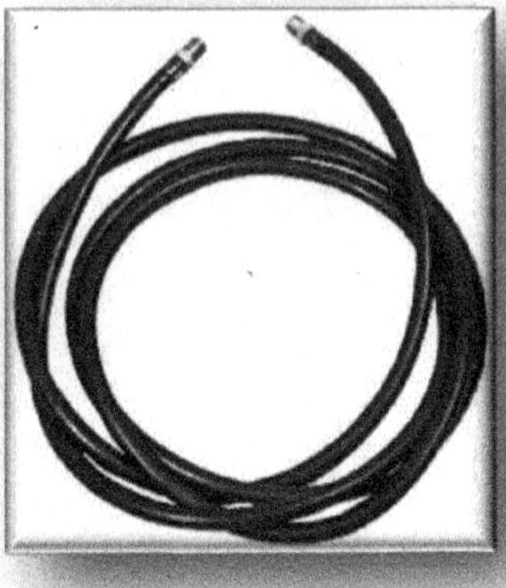

These agencies approve each air-line respirator as an entire unit, including face piece, regulator, and airline. No unapproved respirators may be used at any time. The maximumairline length for any approved respirator may not exceed 300 feet; the maximum inlet pressure at the manifold cannot exceed 125 psi. Any alteration of the respirator or its subassemblies voids the approval. Accordingly, replacement parts must be supplied by the manufacturer for their respirators only.

Additional Information

The following items are provided merely as suggestions for work practices when using air-supplied respirator systems. Most importantly, the contractor should become fully knowledgeable on the use, limitations, and maintenance of the system. The manufacturer's representative will often provide assistance in this area. It is also recommended that an industrial hygienist or other person familiar with such systems be retained for advice during initial setup and until the job superintendent is comfortable with system operations.

The air-supplied mask chosen by the contractor is often one that contains a HEPA escape cartridge which permits the worker to disconnect from the air line in the event of a loss of air pressure or an emergency. While the protection factor drops when on the HEPA filter, the worker normally does not need the high protection of the air line when escaping from the work area.

Plastic caps or tape should be used to cover all exposed airline connections when not in use. This will keep asbestos debris from becoming caked in the quick disconnects. The airlines and manifolds will need to be cleaned completely at the end of the project. It is a good idea to keep the manifold draped with plastic during gross removal and cleanup. Also, coiling the clean air lines and sealing them in plastic bags to be opened inside the next work area is a good practice and saves time.

Thorough training of the asbestos abatement workers is necessary. This becomes even more important when air-line respirators are used. Once workers become accustomed to the air-line respirators, they usually find them much more comfortable than the cartridge respirators since a cool, fresh supply of air is continuously fed into the respirator. This keeps the face piece from fogging and helps reduce the incidence of heat stress, if applicable.

Objectives:

To provide instructions and guidelines to course participants for establishing an ongoing medical surveillance program for employees exposed to airborne asbestos fibers.

Learning Tasks:

Information in this chaptershould enable participants to:

- Understand the need for an ongoing medical surveillance program for workers exposed to asbestos to ensure the safety and health of employees.
- Understand the various elements that comprise an acceptable medical surveillance program.
- Be knowledgeable of the OSHA standards regarding respirator use.
- Understand how the medical monitoring should be conducted, what tests should be performed and why, and what the results of these tests mean.
- Understand procedures for maintaining appropriate records on each employee.

THE IMPORTANCE OF MEDICAL SURVEILLANCE

It is important for all asbestos abatement contractors to establish an ongoing medical surveillance program for several reasons. The three major areas of concern are:

1. The safety and health of all workers;
2. Regulatory requirements; and
3. Other legal liability concerns.

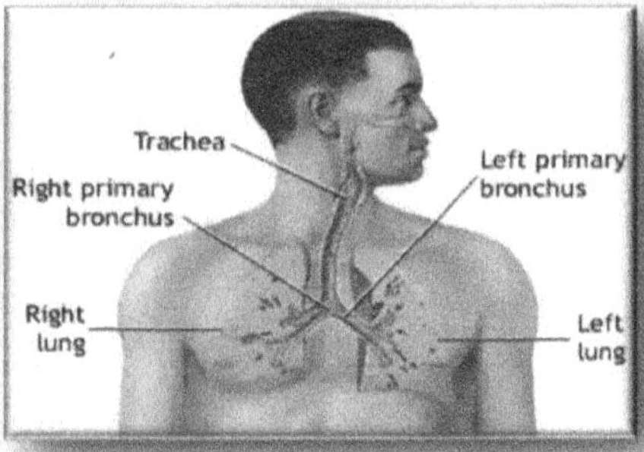

Through implementation of a sound medical surveillance program, an abatement contractor will be able to

1. Verify every employee's medical status at a particular time;
2. Comply with OSHA standards on medical surveillance of workers exposed to asbestos;
3. Reduce other possible liability risks.

In this chapter these three concerns are addressed, in addition to several other considerations associated with medical surveillance programs.

Who Needs Medical Surveillance?

Because of the increased public awareness concerning the hazards associated with exposure to airborne asbestos fibers and because of various regulatory requirements, employers and building owners are finding themselves in situations where they must provide for regular and periodic medical surveillance for their employees. Asbestos-abatement contractors are required [29 CFR 1926.1101 (m)] to provide a medical surveillance program for their employees if they engage in Class I, II and III work for thirty or more days per year; if they are exposed at or above the PEL or EL; or if they wear negative-pressure respirators. For these employees, a medical surveillance program is used for determining their baseline health status (health status before beginning work), monitoring their health for the duration of their employment/project, and providing documentation of their health status, along with their work history, upon completion of their employment/project.

Other employees, such as custodial and maintenance workers, who may encounter and/or disturb asbestos-containing materials while performing their normal duties, should be provided medical surveillance. Examples of these duties might include working above false ceilings with asbestos-containing insulation, installing ceiling tiles, or performing maintenance on pipes or boilers that have asbestos-containing insulation on them. In addition, any employee who wears a negative-pressure respirator as a routine part of his/her job must be medically evaluated on a regular basis (OSHA medical screening is required prior to the use of any respirator). This medical surveillance is to ensure that the use of the respirator does not adversely affect the employee's health.

OSHA Standards – Medical Surveillance

According to the OSHA asbestos standards—29 CFR 1910.1001 for general industry 1915.1001 for shipyard employment, and 29 CFR 1926.1101 for construction and abatement workers—the employer/building owner must provide, at his/her own expense, medical examinations relative to their employees' exposure to asbestos. An acceptable medical surveillance program must include pre-placement and annual examinations unless sufficient evidence is provided demonstrating that an employee has been examined in accordance with the standard within the past one-year period. This standard also outlines the requirements for maintaining medical records on each employee. (Termination examinations are only required by the 1910.1001 standard.)

Pre-Placement Exams

According to the OSHA standards, the required pre-placement examinations must take place before the employee starts the asbestos job. These examinations include:

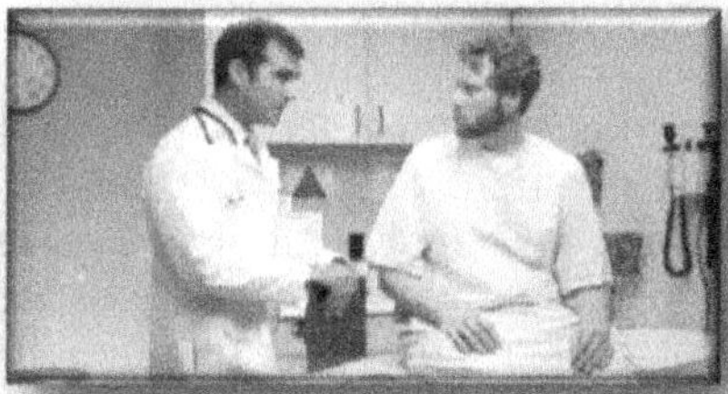

- a comprehensive medical evaluation;
- a medical questionnaire/history to determine the presence of any possible respiratory diseases;
- pulmonary function tests including forced vital capacity (FVC) – (the maximum amount of air that can be expired from the lung after full inhalation), and forced expiratory volume at one second ($FEV_{1.0}$) – (the amount of air forcibly expired in one second after full inhalation).

A chest X-ray (posterior-anterior 14 x 17 inches) is optional at the discretion of the physician; however, it is strongly recommended for the initial examination in order to establish baseline medical data for the employee.

The results of this examination will be used for determining the employee's baseline health status, as well as determining whether an employee is capable of safely working under the requirements set forth by the employer. A physician's report will then be furnished to the employer for his/her files. The physician must provide to the employee a statement that the employee has been informed by the physician of the increased risk of lung cancer attributable to the combined effects of smoking and working with asbestos and the results of the medical examination. Also, the physician is not to reveal in the written opinion given to the employer specific findings or diagnoses unrelated to occupational exposure to asbestos. The employer must provide a copy of the physician's written opinion to the affected employee within thirty days from its receipt. It is very important for the employer to maintain the results of the examination on file for the duration of employment plus thirty years. In the event an employee files suit claiming a disability at some future date, the employer will be able to check his/her records for documentation when investigating whether or not the condition could have occurred as a result of employment with the company.

In addition to the medical reports, the employer must request that the physician provide a statement indicating whether or not an employee is capable of wearing a respirator. This statement should make reference to any lung restrictions that would prevent respirator usage as well as any other limitations associated with its use.

Annual Examinations

According to OSHA 29 CFR 1910.1001, subpart (L)(3) for general industry and OSHA 29 CFR 1926.1101 (M)(2), every employer must provide, or make available, comprehensive medical evaluations to each of their employees engaged in occupations that cause exposure to airborne asbestos fibers (i.e., abatement workers, maintenance people, etc.). Such annual examinations must include, at a minimum:

- completion of a periodic (abbreviated) medical questionnaire;
- a physical examination;
- a study for determining the presence of any respiratory diseases;
- a pulmonary function test that includes FVC and $FEV_{1.0}$.

A chest x-ray (posterior-anterior 14 x 17 inches) would only be required under the general industry standard, to be given at intervals as outlined in the OSHA standard.

The physician is able to compare the annual examinations with the pre-placement evaluations to determine if any changes in an employee's health status have occurred.

If noticeable changes have occurred, the employer and the employee should both be notified, since the situation may require immediate action such as transferring to another job, discontinuing respirator use, or instituting other procedures.

Annual examinations are required for all workers enlisted in a medical surveillance program. With the exception of an abbreviated annual questionnaire, tests to be performed should be the same unless the physician deems other tests necessary for a complete evaluation. Temporary workers should be encouraged to obtain and preserve copies of their medical test results, as annual exams are not required if adequate records document that the worker has been examined in accordance with the construction standard within the past 1-year period.

PHYSICIAN'S WRITTEN OPINION--RECOMMENDED FORM

The following individual has been examined in accordance with the Occupational Safety and Health Administration (OSHA) Asbestos Construction Standard, 29 CFR 1926.1101:

Name of Individual: ______________________________

Birth Date of Individual: ______________________________

Date of Examination: ______________________________

Name of Examining Physician: ______________________________

Name of Medical Facility: ______________________________

Address of Medical Facility: ______________________________

Phone No. of Medical Facility: ____________________

In accordance with the requirements of Chapter(m) of the OSHA Asbestos Construction

Standard, 29 CFR 1926.1101, the examining physician will provide the employer with a written opinion which shall contain the following: (***the physician must check for each statement below)***

1. Based on the results of the medical examination, I have determined this individual

--( ) may ( ) may not --use a respirator while performing asbestos work. Recommended limitations for respirator use:

____

2. The results of my examination today—

   ( ) have ( ) have not--detected a medical condition which would place the employee at an increased risk of material health impairment from exposure to asbestos.

3. ( ) I have informed the above named individual of the results of the medical examination and of any medical conditions that may result from asbestos exposure.

4. ( ) The employee has been advised of the increased risk of lung cancer attributable to the combined effect of smoking and asbestos exposure.

Signature of Examining Physician:

____

Date_________________________

Termination of Employment Examination

Within thirty calendar days before or after the termination of an employee covered by the OSHA general industry standard for asbestos (but not asbestos abatement/construction standard), OSHA requires that each employee exposed to asbestos receive a medical examination. This examination must entail the same items as the annual exam:

- a history for determining the presence of any respiratory diseases:
- pulmonary function testing that includes FVC and $FEV_{1.0}$.
- a chest x-ray as outlined in the OSHA standard 1910.1001 (1)(4).

Records of these exams must be retained by the employer for a minimum period of thirty years to provide documentation of the health status of the employee. This thirty-year period is necessary because the latency period associated with asbestos-related diseases often ranges between 15 to 30 years. Thus, if an employee files a claim 25 years later, the employer will have records on file for reference.

REASONS FOR SPECIFIC TESTS

All of the tests performed during preplacement, annual, and termination medical examinations are required so that the human body systems that are most likely to be affected by exposure to elevated levels of airborne asbestos fibers can be properly evaluated by trained medical personnel. Some specific reasons for each test are discussed below.

Pulmonary History

This part of the examination is simply a questionnaire (contained in the appendices of the OSHA asbestos standards) that is completed by the employee and physician. This questionnaire identifies the potential for respiratory diseases. Several questions relate to chronic lung diseases, while other questions address the employee's personal habits, such as smoking. Smoking is often of particular concern for workers who may be exposed to asbestos because smoking is known to compound or intensify the effect of asbestos on the lungs. Studies indicate that an asbestos worker who smokes is at least 50 times more likely to develop lung cancer than nonsmokers who do not work with asbestos.

Physical Examination

Criteria to be evaluated as part of the routine physical examination often include a medical history, blood pressure, pulse, vision (depth perception, peripheral vision), an audiogram (hearing test), urinalysis, complete blood count (CBC), and follow-up classification with appropriate recommendations.

Pulmonary Function Tests

These tests are conducted to determine if a person's lungs are expanding normally, and if adequate air movement in and out of the lungs is occurring. A spirometer is used for conducting the FVC and $FEV_{1.0}$. tests. If the $FEV_{1.0}$. is reduced, a possible obstruction or problem in an employee's lungs may be present. If the FVC or the ratio of $FEV_{1.0}$ to FVC is reduced, restrictive changes in the employee's lungs may have taken place.

Chest X-ray

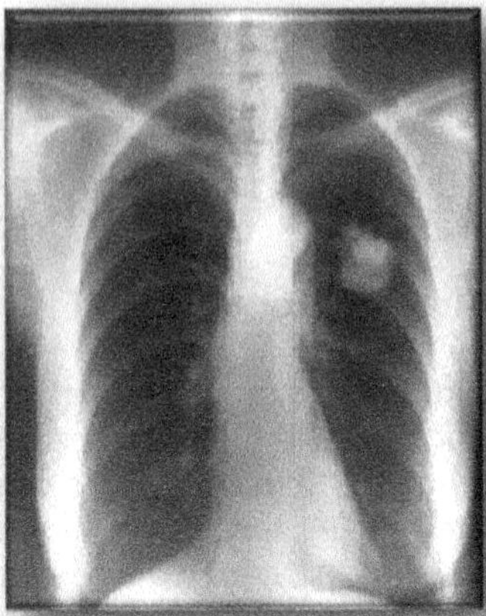

X-rays (posterior-anterior 14 x 17 inches) are performed primarily to detect irregularities in the lungs or the heart, including any fibrosis or pleural plaques induced by a person's exposure to asbestos. Chest x-rays may be used as a baseline for comparing future x-rays.

Chest x-rays should be interpreted by a certified "B" Reader, a physician (often a radiologist, occupational medicine physician, or pulmonologist) who has received specialized training in the interpretation of chest x- rays, specifically relating to occupational lung diseases. 'B Readers' are required to pass a proficiency test administered by the National Institute of Occupational Safety and Health (NIOSH) in Morgantown, West Virginia. NIOSH is a federal agency under the Centers for Disease Control and Prevention.

An individual over 40 years of age or who is otherwise at increased risk when working with asbestos and when complying with OSHA asbestos standards should have, as part of their physical examination, an electrocardiogram (EKG). The use of a respirator places an increased strain on the heart; for individuals with heart disease, appropriate actions should be taken (e.g., transfer to a job that does not require respirator use).

Costs Associated with Medical Surveillance

The costs of employee medical surveillance examinations performed at occupational medical clinics are relatively inexpensive compared to the cost of private physician's fees. In some cases, it may be possible to obtain group discounts if enough employees are involved.

Check with your instructors on the current prices of the following:

- Chest x-ray
- PA/lateral
- 'B" Reader Fee
- Pulmonary Function
- Physical Examination
  (pulmonary history included)
- Electrocardiogram
- Gastro-intestinal Hemoccult

SUMMARY

Important information to obtain from this chapterof the course includes an understanding of why a good medical-surveillance program is essential for employers to ensure the safety and health of their employees, and also to reduce their liability potential for claims pertaining to asbestos exposure.

Also, it is important to have a firm understanding of the OSHA requirements regarding medical surveillance programs for employees exposed to more than 0.1 f/cc of airborne asbestos determined by an 8-hour TWA sample, more than 1.0 f/cc determined by a 30-minute sample or other employees who must routinely wear respirators as a part of their job. Additionally, it is important to understand the reasons associated with each of the specific tests that comprise an acceptable medical evaluation program.

Objectives: To review the primary tasks that should be performed by the contractor before beginning an asbestos abatement project. Also, to discuss important considerations that should be examined before agreeing to the job specifications and negotiating a price.

Learning Tasks: Information in this chapter should enable participants to:

- Understand what to look for when conducting the pre-bid walkthrough survey of the job site.
- Ask specific questions relating to job site conditions and how they affect agreements under the specifications.
- Become knowledgeable of good techniques for screening and selecting abatement employees.
- Understand the type of information that needs to be covered in a training program for abatement employees.
- Understand the design and use of a project logbook and its importance during an asbestos abatement job.

ASSESSING THE WORK AREA

Note: Various states have specific requirements for work area preparation and abatement work practices. See your instructor for state requirements.

An important "rule of thumb" for any asbestos abatement contractor is to never bid or accept a project without first viewing and assessing the site. There is much valuable information to be gained during one of these assessments, such as determining the size of the job (number of square feet or linear feet of asbestos- containing material) or examining the configuration of the ceiling surface noting the accessibility for abatement of the material (irregular ceiling shape can increase the amount of asbestos-containing material originally believed to be present). A survey such as this also provides a basis upon which the contractor can formulate an effective strategy for asbestos removal and/or control. Existing damage or potential problems (expensive rugs, light fixtures, etc.) need to be identified and procedures for resolution dealt with before work begins. Some of the important items an abatement contractor will need to look for on this pre-bid survey are outlined in this chapter.

Check Analytical Results of Bulk Samples

The first questions that a contractor probably should ask during the pre-bid walkthrough survey are who did the initial survey to identify the asbestos, what type of sampling was conducted, and what forms of analysis were used. The contractor should ensure that appropriate bulk sampling was performed by qualified individuals using proper analytical methods. Any inspections conducted in a School (K-12) or public and commercial building fall under the AHERA and ASHARA requirements respectively (see "Regulations" chapter) and must have been completed by an accredited inspector. The AHERA regulations require that bulk sample analysis be done by laboratories accredited by the National Institute of Standards and Technology (NIST) National Voluntary Laboratory Accreditation Program (NVLAP). Any bulk samples collected under the National Emission Standards for Hazardous Air Pollutants (NESHAPS) must be analyzed using the point counting method under certain conditions, and the reader is referred to the chapter of the notebook discussing asbestos regulations. The contractor should then review the analytical results of the bulk samples to determine the types and percentages of asbestos present.

There are several reasons why this type of information will be of benefit to the contractor. First, the analytical reports provide excellent documentation that can be used in establishing a project file. This file can then be used as a good source of reference should any questions arise concerning the asbestos-containing materials in the building. Information contained in the analytical reports is also important because different types of asbestos will require various handling techniques. For instance, amosite is considered by some scientists to be more hazardous than chrysotile, in addition to not accepting wetting agents as well, and will require different handling procedures.

Fiber counts will usually be much higher when handling amosite as opposed to chrysotile. If analytical reports are not available prior to or during the survey, the contractor should obtain his/her own by including it as part of the assessment. It is important that the information from these reports be used as the main criteria on which to base decisions, rather than word-of-mouth from a resident maintenance worker or other building occupant which could lead to confused facts or other misinformation.

Inspect the Nature of the Asbestos-Containing Material

The contractor should determine the hardness and texture of the asbestos-containing material to be removed (must touch it). He/she should also note whether or not it has been painted. (Note: A high efficiency (HEPA) cartridge type respirator should be worn when conducting these tests.) The contractor may also wish to test a sample area of asbestos-containing material to determine its ability to absorb amended water. This can be done by using a plant sprayer. If the material cannot absorb wetting agents, other appropriate strategies will need to be developed which may increase the cost and project time (see Chapter XI, "Confining and Minimizing Airborne Fibers").

Check Accessibility of Material

Note the accessibility of all materials for removal, that is, whether or not the asbestos-containing material is accessible enough to remove. If not, an alternative means of control might have to be used such as encapsulation or enclosure. Several factors that may enter into this determination are ceiling height, false ceilings, pipes, sprinklers, ducts, sloping floors, fixed barriers, etc. The building access may also require innovative methods. All of these considerations will affect the bidding price.

Check for Difficulty of Isolating the Work Area

Another important concern is isolating the area in which removal will take place. Is it possible to enclose the area completely by using polyethylene? Or, will other measures have to be implemented in certain areas to adequately isolate the removal site? In cases such as school buildings, it may be easiest to simply line the walls and floors with two layers of polyethylene since the contractor will usually remove all desks and chairs from the work area. However, in cases such as a church, computer room, or industrial facility, plywood and plastic enclosures may have to be constructed so that the materials left in the room will not be contaminated by the asbestos removal activities or damaged by water. Another chapter, "Preparing the Work Area and Establishing the Decontamination Unit," further discusses these practices.

Determine if Areas Adjacent to Abatement Activity Will Be Occupied

If areas adjacent to the abatement activity will remain occupied, several important practices should be observed. Most importantly, the HVAC system will need to be altered, or the opening of the duct into the work area should be completely sealed off. If possible, the supply, return, and exhaust systems serving a specific work area should be shut down or disconnected. Sealing of the HVAC helps ensure that airborne fibers will not be drawn into the air return system and dispersed throughout adjacent areas, or the supply system will not place the work area under positive pressure and cause airborne fibers to escape. An excess number of negative pressure units are a prudent method to minimize the risk of loss of negative pressure and ensuing fiber release to the occupied areas. To provide documentation that contamination of adjacent areas has not occurred, a qualified person should take background air samples in each of the areas before abatement work begins. These results are then compared to the results of samples taken in these areas during and after the work is completed. By doing this sampling, it can be demonstrated that other areas were not contaminated as a result of the asbestos abatement work.

Determine Room Volume and Natural Air Movement in the Work Area

During this walk-through survey, consideration should be given to the number and placement of negative air units. A calculation of the air volume in the work area is necessary for determining the number of units needed to achieve the desired number of air changes per hour. Also, the way in which air will move through the work area is a consideration in placement of the negative air units. This idea is further outlined in the chapter entitled, "Confining and Minimizing Airborne Fibers."

Check Items Requiring Special Protection

During the pre-bid walkthrough, items requiring special protection should be noted. These items might include wood paneling, trophy cabinets, glass piping, carpets, lab equipment, dangerous chemicals, computers, and elevators.

In the case of wood paneling, common sense should be used when hanging polyethylene to enclose the work area. Care must be used when tacking up the plastic so that the paneling will not become damaged. Nails should be placed between the panel strips in the natural gaps as near the ceiling as possible to prevent any small holes from being visible.

For trophy/storage cabinets that are stationary and must remain in the work area while removal is taking place, proper measures must be taken to ensure that the cabinet is adequately enclosed with six mil polyethylene.

During this initial survey, the contractor should note the condition of any of these cabinets, and the exact contents of each to prevent any future conflicts that could result if someone were to claim that something was damaged or missing.

Glass piping is another item that the contractor should note during the pre-bid walkthrough since special procedures must be followed to ensure that it does not become damaged. These glass/ceramic pipes will often contain hazardous materials, (i.e., acids, hazardous waste, etc.). Therefore, the pipes should be tagged and/or labeled as containing hazardous materials, and workers should avoid contacting them if possible. These glass pipes are often found in the vicinity of other pipes that have asbestos-containing lagging on them. Therefore, contingency procedures must be established to prevent and handle hazards that could develop from working around these pipes.

Determine Contractor's Responsibility Concerning Moveable Objects

During the pre-bid walkthrough, it must be determined what the contractor's responsibilities are concerning any and all moveable objects.

Discuss with the owner what equipment, furnishings and finishes (i.e., light fixtures) must be cleaned, moved, stored and reinstalled or discarded and replaced. This one area, if neglected, can be quite costly to the contractor or the owner. Are there objects which need to be cleaned before moving? Where are these items to be stored? Will the owner provide space in an area not being abated or must the contractor provide space (i.e., truck or rented storage space)? Is reinstallation also, then, the responsibility of the contractor? Is it more cost effective to discard some items (i.e., carpeting, light fixtures, contaminated fabric, furniture, drapes) and replace? Some items moved and stored will, inevitably, be damaged or broken. Is the contractor responsible for replacement costs? Having all these considerations addressed during the pre-bid walkthrough will help alleviate misunderstandings and increased costs for both parties once the job has been bid.

Special note should also be made of where carpeting is located in the facility. In most cases, the carpeting should be removed completely from the area in which the asbestos removal will be taking place. When fibers settle on a carpeted surface, they often penetrate through to the floor and become trapped underneath. Once this occurs, repeated traffic over the area will cause the fibers to be redistributed throughout the surrounding air. If carpet is specified for removal, assess the difficulty of removing it (i.e., the carpet may be glued in place). Also, consideration must be given to disposal requirements/procedures.

Note Any Materials or Equipment That Will Require Special Handling

Additionally, lab equipment and/or dangerous chemicals should be examined closely by the contractor during the pre-bid walkthrough survey. It may be necessary to remove much of the equipment and/or chemicals from the work area before abatement activities take place. If the contractor's employees will be moving expensive lab equipment or chemicals, the contractor should ensure that all items are appropriately handled through training and/or direct supervision. This may be a tedious process requiring extra time to complete. In some cases, the building owner may have his/her own maintenance personnel perform these functions before the contractor comes in to begin work.

Note Stationary Objects That Require Special Attention

As previously mentioned, if the abatement work area will be in a room that contains computers that cannot be moved, other strategies must be developed such as building an elevated platform (plywood and plastic) over the terminals.

Elevators can also be a major problem on an asbestos abatement job. The elevator or the shaft can become contaminated with asbestos-containing materials, or their movement can cause air displacement in contaminated areas. The contractor will need to take special precautions to properly seal off the doors with six mil polyethylene (even plywood in some cases) and to key the elevator not to stop at the floor(s) on which the work area is located.

Utilities

To prevent any misunderstandings or conflicts, it is imperative that the job specifications spell out exactly who is to pay for the utilities used during the project. Usually, the building owner will pay these expenses, but if not, this should be clearly understood by both sides before work begins. Also, does the building have existing electrical and water service, or must the contractor provide these services? Likewise the waste water filtration and disposal method should be agreed upon and specified (see the "Waste Disposal Requirements" chapter).

Existing Damage

The contractor should also document *all* existing damage in the areas in which his/her employees will be working. Photographs, videotapes, diagrams, lists, and tape recordings may be used for this purpose. This documentation should include all surface damage (walls, tables, desks, etc.), vandalism, roof leaks, or other water damage. This consideration is important because often after a project has been completed, the building owner, or another facility operator will claim that some damage occurred as a result of the contractor's work.

By using the list that was developed at the beginning of the project, the contractor can verify whether the damages were existing and not a result of the contractor's work.

Safety Considerations

Additional safety hazards that need to be considered include all electrical circuits and/or receptacles, equipment, etc. Since the work are is an asbestos abatement job will commonly contain large amounts of water, the potential for electrical hazards will be greatly increased. During the pre-bid walkthrough, the contractor should make note of all these potential hazards. Once the building owner is made aware of these situations, an appropriate plan of action can be implemented. It may be possible and appropriate to shut down all power to the work area while the project is going on. If not, other precautions will need to be taken. This is discussed in greater detail in the chapter entitled, "Safety and Health Considerations Other than Asbestos."

Also, a major area of concern when assessing a facility prior to beginning work is identification of any hot surfaces (pipes) that could present a hazard to abatement workers. First, it should be noted whether the pipes will be active or inactive. If they are active, appropriate measures will have to be taken to ensure that the workers will not contact these surfaces. If the lines are inactive, work may be carried out as it would on any other surface of normal temperature. The contractor should investigate the types of re-insulation that will be required on surfaces and pipes after the asbestos-containing material has been removed. The original material was there for some specific purpose, thus, a replacement material with similar properties will probably be necessary.

Work Area Configuration

Another area of concern during the walkthrough should be the configuration of the walls and surfaces for attaching tape. This is important to determine how the polyethylene sheeting will have to be hung to adequately enclose the work area. Care must be used when hanging polyethylene so the walls will not be damaged, but the plastic will remain in place until intentionally moved. This is an important consideration at this time since the contractor will have to estimate how much material will be needed to enclose all work areas. Many times, the building owner may want the project to be inconspicuous to the general public. Therefore, opaque polyethylene may have to be used to construct tunnels from the work areas outside to the waste disposal trucks. Additionally, depending on the nature of the work area, special tools, equipment, and man lifts or scissor lifts may have to be used during prepping of the work area.

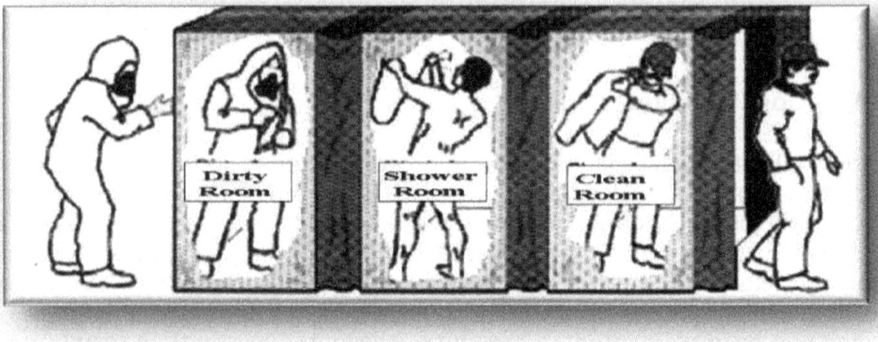

The location and type of decontamination the units should also be a major consideration before submitting a bid. Will it be possible to have one central decontamination unit, or will it be necessary to establish multiple stations? Some contractors may have their own units (i.e., trailers), but many choose to build them on-site. Many buildings in which asbestos removal takes place already contain shower facilities (i.e., school buildings, gymnasiums, etc.). Under no circumstances should the contractor ever permit his/her employees to use these as part of the decontamination sequence. Separate facilities should be constructed using appropriate waste water filtration equipment. An advantage of building temporary site units is that the chance of residual contamination is reduced since they will be demolished and disposed of at the end of the project.

Air Monitoring

Another important aspect that must be considered by a contractor before bidding a project is who will pay for the air monitoring, and whether or not the person conducting the monitoring is qualified. This should be established in the specifications. The building owner should always be responsible for the daily air sampling, but the contractor is often responsible (and required by OSHA) to conduct personal air sampling on the asbestos abatement employees. (Note: This is discussed in greater detail in the "Air Monitoring" chapter.)

Time

The contractor should ensure that the job specifications allow adequate time for their company to complete the job with a high degree of quality. If specifications call for a "hurry up" job, the contractor should inform the building owner or architect/engineer if they do not feel that adequate time is available to complete the project.

Attempting to perform the job hastily may only result in sloppy work and may needlessly endanger the health and safety of employees or other building occupants.

These are not <u>all</u> of the special considerations that need to be examined when conducting a pre-bid walkthrough survey of an asbestos-containing facility; rather, there are some common concerns that should typically be investigated before beginning any asbestos abatement project. They are important because they could cost the contractor's company a substantial amount of time and money, in addition to possibly endangering the lives of employees or other building occupants. It is imperative that the contractor and the building owner have a firm understanding as to exactly how each step of the project will be carried out.

MEDICAL SURVEILLANCE

Contractors are required as employers to provide, at no charge to the employee (if exposed to asbestos), a physical examination by a qualified physician. There are specific items that these physicals must entail (refer to OSHA 29 CFR 1926.1101). The contractor may use the results of these physicals to screen potential employees that may have had previous exposures to asbestos. If possible, the contractor should avoid hiring a heavy smoker as a removal worker, or anyone else that would naturally be at an increased risk from previous exposure. This topic is further detailed in the chapter entitled, "Establishing a Medical Surveillance Program."

EMPLOYEE TRAINING

Any workers who will be in or around an asbestos abatement work area should, as a minimum, be advised of the hazards associated with asbestos exposure, be trained in how to adequately protect themselves from exposure during the course of the project, and be trained in correct job procedures for each of their positions. Many states now require training for all asbestos workers and supervisors. The Asbestos Hazard Emergency Response Act (AHERA) of 1986 and the Asbestos School Hazard Abatement Reauthorization Act (ASHARA) of 1990 require that all response action workers and supervisors attend a U.S. EPA approved (and state approved, when applicable) training course and pass an exam. OSHA also has training requirements for individuals based on the type of work that will be performed, and requires training for the use of respirators. Incorporated into this training program should be a thorough explanation of all mandatory regulations governing asbestos abatement. A good way of documenting that this training has taken place is to develop a formal training session at which attendance is a mandatory condition of employment.

Although state or federal training may not be required for some projects, it is still highly recommended in order to provide informed and trained personnel workers within the asbestos abatement field.

Background Information

A good, in-depth training program should cover many concepts dealing with the various aspects of asbestos abatement projects. Employees should be given background information on asbestos, such as what it is and where it comes from. Also, they should be informed as to how asbestos was used and why.

An architect, engineer, or someone else with a good technical background may be best suited to present this part of the training program.

Asbestos Health Effects

The next phase of the training should be an outline of the dangers or health hazards associated with exposure to asbestos fibers. Someone with a good understanding of the medical hazards associated with breathing asbestos should give this part of the training session. It is important that not only the health hazards be discussed, but also how the fibers enter the body, and what happens once they are contained inside the lungs. Fiber size, visibility, and settling times are all important information. A film or slides may be helpful in illustrating these points.

Personal Protective Equipment

After employees are made aware of the health hazards associated with asbestos exposure, the next phase of training should be on what they can do to protect themselves from this exposure (i.e. work practices and personal protective equipment). This training should include step-by-step instruction on how to perform each task associated with their jobs (i.e. glovebagging, wetting and scraping, etc.). Also, training should include a comprehensive review of the use of respiratory protection, including the following aspects:

- How to put on and take off the respirator
- Cleaning and maintaining respirators
- Inspection of respirators
- Fit testing of respirators
- Discussion on uses and limitations of different types of respirators
- Hands-on experience (look at various parts)

Note: The training requirements of an effective respiratory protection program are addressed in the chapter entitled, "Protecting the Worker: Respirators and Protective Clothing."

It is also important that workers be properly trained in the use of protective clothing. They should be made aware of its limitations, and how it should be used to optimize the protection factor. (Note: See chapter entitled, "Protecting the Worker: Respirators and Protective Clothing.")

Asbestos Regulations

The next phase of the training program is a discussion of of all applicable EPA and OSHA regulations regarding asbestos abatement projects. Also, there may be certain state or local regulations of which employees need to be aware. This part of the program should not be extremely detailed, rather it should provide the employees with a good understanding the of what they should or should not do when conducting removal of asbestos-containing materials. It should be emphasized that the main concern is the safety and health of the workers, rather than simply the concern of receiving a citation for a violation.

Work Area Construction

The fifth phase of employee training should deal with proper techniques for sealing off the work area. In this chapter, employees will be instructed on what to look for before sealing off the work area and also how to construct a safe and effective enclosure. Employees should first be made aware of what an HVAC system is and how it affects the air movement through an area. They should also be instructed on how to shut down the system and seal off outlets and inlets so that airborne fibers will not be drawn into it. Employees should then be instructed in proper techniques for erecting plastic barriers, draping the walls, floors, and furniture with polyethylene. This also includes construction of airlocks and change rooms, in addition to posting appropriate warning signs, etc. Also, it is important to inform employees that if a puncture develops in the polyethylene enclosure while the work area is active, they should stop work and immediately seal the leak.

Minimizing Asbestos Fiber Generation

Following the session on sealing off the work area, workers should be trained in how to effectively confine and minimize airborne fiber generation. This can best be accomplished through proper use of wet methods (i.e., spray the asbestos- containing material with amended water). Workers should also be informed at this time that different forms of asbestos will react differently to the application of water. For example, chrysotile will typically accept water, while amosite is generally more resistant to wetting.

Therefore, employees will have to take appropriate protective measures since airborne fiber concentrations will be potentially higher when a removal job involves amosite. Employees should be instructed in methods of misting the air with water, and also in the proper methods of using the HEPA vacuum. Additionally, the function of negative air units should be outlined and employees made aware of the need to ensure that these units are kept running so that if a rupture occurred in the enclosure, fiber leakage would be minimized.

Safety and Health Issues

A very important aspect of employee training that is often taken lightly is the recognition and control of safety and health hazards (other than asbestos) in an asbestos abatement work area. Proper training can help reduce employee injuries, and lost time accidents. Subject areas that should be covered in this part of the session include the proper use of scaffolding, how to recognize and/or eliminate trip/slip hazards, the proper use of ladders, the identification of any electrical hazards, the identification of any chemical hazards, and how to avoid heat stress/heat stroke situations.

Work Area Cleanup

The next phase of employee training should entail cleaning up the work area. This cleaning will take place after gross removal has occurred and all residual debris is ready to be disposed of. Wet cleaning techniques should be reviewed (wetting the waste and collecting it off the floor). Settling times should also be discussed.

Waste Disposal

Correct disposal of asbestos-containing debris is also an important aspect of an abatement employee training program (specifically for employees who will be directly involved with disposal operations). This part of the program should include discussions on the need to place the wetted waste in appropriately labeled six mil polyethylene bags. These bags should then be double bagged before being loaded into the enclosed truck to be taken to the landfill.

It is important that any employees who might be involved with this type of operation in some way be made aware of the proper procedures for carrying out these waste disposal activities and the protective equipment required.

Work Area Inspections

Another important aspect that an employee training program should include is information on final inspections/air sampling and why it is important. The reason that employees need to be aware of what the final inspections will entail is that when they finish work in a certain area, they can conduct a fairly thorough visual inspection themselves.

Air Monitoring

Employees should also be informed as to why air sampling is being conducted and what the results mean. Employees should be informed that they may be asked to wear a personal air sampling pump while they are performing their job so that the fiber levels they are exposed to can be closely monitored. They should be requested to cooperate with the industrial hygienist when it is their turn to wear the sampling equipment. It must also be emphasized to the employees not to tamper with the sampling equipment they are wearing since the results will indicate the level of airborne fibers to which they are being exposed.

Worker Decontamination

One area that the training program should cover is an explanation of the personnel decontamination sequence. This should cover procedures to be followed when beginning or finishing a shift of asbestos abatement work. When beginning a shift, employees should be instructed to enter the clean room first, put on their protective equipment/clothing, and proceed through the shower area and into the dirty equipment room before entering the work area. When finishing a shift, employees should be instructed to enter the dirty equipment room first, remove all of their protective clothing (except respirators), and then take showers. Respirators should be removed and washed while the employee is in the shower. After the employees complete the shower, they can then go to the clean room to change back into their street clothes. Employees should also be instructed that in emergency situations, the emergency will probably override the potential of adjacent area contamination, and good judgment should be used if someone needs to get out of the work enclosure very quickly (employee has a heart attack, fire, etc.). Emergency break-through points in the polyethylene enclosure should be clearly marked so that they will be easily accessible.

Hands-On Training

Once all of this formal classroom training is completed, ample time should be provided for employees to participate in hands-on training or workshops.

Demonstration in these workshops should include proper techniques for glovebagging, wet and scrape methods, constructing work area enclosures, personal protective equipment, etc. This will be the most effective way to illustrate exactly how these typical asbestos abatement procedures should be conducted.

Throughout the training program, slides, video tapes, and handouts should be used when possible. A mixture of training techniques results in better learning. Also, employees with prior experience in asbestos abatement may have valuable input during this session. These comments should be encouraged by the instructor and any misinformation should be immediately corrected without "putting the person down." It is important for the instructors not to get overly technical in any one area. At the end of the program, the written test should be administered. Results of these tests should be used to spot areas where employees may need further training. Also, at this time, employees should sign a form indicating that they have received training. The tests and the forms should then be placed in a file so that there will be documentation that employees were trained appropriately. Periodic retraining should be a part of each employee's duties and the documentation maintained.

DESIGN AND USE OF A PROJECT LOGBOOK

Prior to the start of any asbestos abatement project, a logbook should be established. This logbook will serve as a vehicle for maintaining all the records associated with a project. At a minimum, included in the book should be copies of the employee's medical reports, copies of any accident, injury, or accident reports, unusual events or occurrences (such as failure of negative air system or containment barriers), air sampling results, notes concerning any deviation from standard work procedures, sign-in sheets, employee training/certification documentation, and all other pertinent documents, permits, correspondence, photographs, or records. Many of these records will be duplicated elsewhere, such as medical records in the employee's personal file, etc.

The logbook serves many important functions. First of all, it provides an easy reference for projects that can be presented at any time during a project, or long after its completion. It may be reproduced by the contractor to demonstrate to future clients the procedures followed during a project. Likewise, the logbook can be an important tool for planning future jobs and estimating costs. When planning a project similar in nature, it can aid in estimating how long the project will take to complete, how many people will be necessary, and how to approach specific problems. The following examples may help to illustrate this point.

Example: Prior to the start of one job, the contractor probed the depth of the fireproofing and found it to be approximately three inches deep. During the removal, they discovered that approximately fifty percent of the fireproofing was over six inches deep, and in some areas, nine inches deep. A few extra minutes of probing the depth of the fireproofing would have saved much time and money during this project. Unfortunately, this was not the first time that this had happened to this company. Had it been recorded in a logbook the first time it occurred, and changes been made in the standard procedures for estimating the amount of material made, this problem could have been avoided.

Example: A removal project of twenty-four thousand square feet was two days ahead of schedule with only the sprayback of treated cellulose remaining to be completed. Three days after this sprayback material was applied, it began to fall from the ceiling. It took the contractor an extra week of work to remove this material and replace it with a different substance. The problem resulted from the inability of the material to adhere to the substrate since temperatures during application exceeded 95°F in some cases. Notes of this problem were maintained in the project logbook and corresponding instructions added to the standard operating procedures to prevent this from occurring again.

Another function a project logbook may be useful for is possible protection for a contractor regarding liability over specific projects. A logbook indicates that the contractor performing the work actually attempts to do the best job possible using state-of-the-art techniques. The sign-in sheets maintain a record of all people entering and exiting the work area, for what purpose, for how long, and what personal protective equipment they need. This information, coupled with the air sampling data, can quickly be used to estimate how much asbestos the person was exposed to and for how long. Copies of daily inspection reports will also reveal if employees were wearing the appropriate protective equipment and whether or not it was adequate in protecting them from the airborne fiber levels documented by the air sampling results. This information would be very valuable if needed for litigation in the future. It is important to note that all records must be kept, not just a portion of them.

Example: The is a hypothetical example.

> The year is1996; a woman dies of lung cancer. Her husband recalls that 20 years ago, asbestos fireproofing had been stripped from the boiler room of the building in which she worked. A suit is filed against the building owners and the contractor who performed the removal work. Although the contractor performed air sampling throughout the projects, no records were kept regarding work practices, other people in the area, whether the air handling system was on or off, or where the waste was disposed. Since this is a hypothetical case, speculation on the outcome would not be appropriate. However, the contractor would have a better defense if proper records were maintained.

The logbook should be well organized, but in a style decided by the contractor. There are two common methods of organization. First, there is the day-by-day method such as a captain's log. If this method is chosen, a loose-leaf or bound notebook with dividers labeled with each day should be maintained for each job. Be sure to make entries on days that no work is done, including how the integrity of the jobsite was maintained.

Another more common method of organizing a logbook is by activity. Using this method, a loose-leaf notebook is divided into each activity and all documentation, notes, and receipts concerning that activity are maintained in the appropriate chapter.

The following outline is a suggested logbook organization format. (NOTE: This is just an outline; depending on the requirements of each project, some chapters may not apply, while additional ones may be necessary.)

LOGBOOK ORGANIZATION

CHAPTER	CONTENTS
Pre-Work Papers	EPA or state notification forms, any necessary state licenses, county or city permits, (contractor license, disposal permits, etc.). Records regarding the bonding company, size of bond, insurance coverage, etc.
Contract Specifications	Contract specifications, including all drawings/diagrams would be in this chapter.
Personnel	Personnel records including employment application, W-4 withholding forms, medical records, state certification numbers and training certificates (on each worker and supervisor), and any other records pertaining

to each employee. Some firms also have their employees sign certificates stating that they have read and understand the OSHA asbestos standard (29 CFR. 1926. 1101), been trained in asbestos removal techniques, trained and fit tested for respirators, etc.

Sign-In Sheets

A separate chapter containing the daily sign-in sheets indicating when each employee went in and out of the work area, their affiliation, and their purpose for entering the work area. In this chapter would be a list of all authorized personnel permitted to enter the contaminated area. Also in this chapter is a record of each employee's work hours for payroll purposes.

Subcontractors

This chapter would be a record of all subcontractors' activities including copies of the contact, names, dates, etc.

Air Monitoring

All air sampling for the project should be included in this chapter. Area air sampling and personal sampling results should be presented. Also presented in this chapter should be a copy of the sampling and analytical method used along with information concerning who performed the work.

Waste Disposal

Records of waste disposal activities including trip tickets should be kept in this chapter.

Daily Inspection Reports

Copies of daily inspection reports should be maintained. These reports, addressed elsewhere, should also indicate who performed the inspection, date, and time of inspection. It is extremely important to include comments on unusual aspects of the project and to address any problems that arose and how they were handled.

Other Handout Materials

Other handouts materials may be added as necessary. Possibly injury/illness reports, receipts for rental equipment lodging, outside inspections, newspaper clippings, etc.

The responsibility of maintaining the logbook should be assigned to responsible personnel. Normally, this function is performed by the job site supervisor or the other person responsible for coordinating activities at the work site.

Upon conclusion of the job, this person may write a one page summary of the project. This summary can then be compiled with others and produced as evidence of previous jobs performed by the contractor to prospective clients. The two attachments that follow are examples of a daily project log and a sign-in sheet, respectively.

V. DAILY PROJECT LOG

Date: ____________________

Project Name: __________________________________ Job No.: __________

Superintendent: ____

NOTE: Fill in GENERAL comments on routine progress for the above date. DETAIL major problems and action taken, injuries, equipment breakdown, unusual conditions or situations, inspections, hiring or firing of personel and any other occurrence which may affect the project. This log may be used as a legal document.

____

____

____

____

____

____

____

____

SIGNATURE: ____________________________________

JOB SIGN-IN/SIGN-OUT AND VISITORS LOG

DATE ______________________________

All personnel must sign in or out every time they enter/exit work area

PROJECT ______________________________

SUPT. ______________________________

PLEASE SIGN CLEARLY

NAME

SS#

TIME IN

TIME OUT

REASON/PPE ISSUED

EXAMPLE INVENTORY CHECK-LIST FOR ASBESTOS ABATEMENT PROJECTS

CONSTRUCTION MATERIALS

_____Lumber
_____Plywood
_____Nails
_____6 Mil Poly, 12' x 100' Clear*
_____6 Mil Poly, 20' x 100' Clear*
_____6 Mil Poly, 28' x 100' Clear*
_____6 Mil Poly, 32' x 100' Clear*
_____4 Mil Poly, 12' x 100' Clear*
_____4 Mil Poly, 24' x 100' Clear*
_____4 Mil Poly, 20' x 200' Clear*
_____10 Mil Poly, 20' x 100' Clear*
_____6 Mil Poly, 20' x 100' Clear*
_____2" Duct Tape
_____3" Duct Tape
_____3" Black Tape
_____White Tape
_____Tape Guns
_____Regular Scaffold Pieces
_____4" Ladder
_____6" Ladder
_____8" Ladder
_____10" Ladder
_____12" Ladder
_____24" Extension Ladder
_____40" Extension Ladder
_____Tarps
_____Barricade Tape
_____Warning Signs
_____Water Tubs, Large
_____Water Tubs, Small
_____Scaffold Braces
_____Scaffold Pins
_____Scaffold Boards
_____Scaffold Jacks
_____Walkboard 24'
_____Walkboard 20'
_____Wire Nuts
_____Wall Nail Blocks
_____Concrete Block Nails
_____Conduit Frames
_____Straight Conduit
_____Power Saw
_____Flood Lights
_____Spot Lights, Clip on

*(Fire retardant)

DECON MATERIALS

_____Waste Disposal Bags, 6 mil, Labeled
_____Water Filter Housing
_____Water Filter, 100 µm
_____Water Filter, 50 µm
_____Water Filter, 25 µm
_____Water Filter, 5 µm
_____Water Hose
_____Hose Attachments
_____Chains for Lockup
_____Encapsulant
_____Airless Spray Rigs
_____Sump Pumps
_____Teel Pump
_____Wetting Agent
_____Mixing Tank
_____Garbage Cans
_____Drums
_____Sponges
_____Cotton Rags
_____Buckets
_____Shower Pans

PAPERWORK (SUPERVISOR)

_____Project Logbook
_____Regulatory Book
_____Sign In/Out Forms
_____Disposal Forms
_____Worker Release Forms
_____Accident Report Forms
_____Respiratory Training Forms
_____Worker Training/Certification File

NOTE: This is not a complete check-list. Items vary from project to project. You are encouraged to prepare your own check-list.

PERSONAL.PROTECTION	HANDTOOLS (cont.)
_____Safety Boots	____Bent 3"Scraper
_____Rubber Boots	____Straight 3" Scraper
_____Deck Shoes	____12" Wide Knives
____ Protective Clothing, Disposable	____6" Wide Knives
_____Head Covering	____Putty Knives
_____Rubber Gloves	____Hawk Bill Knives
_____Foot Covering	____Wire Cutters
_____Leather Gloves	____Razor Knives
_____Hard Hats	____Razor Knives Blades
_____Hearing Protection	____Measuring Tape
_____Eye Protection	____Nylon Brushes
	____Wetting Wands
RESPIRATORY EQUIPMENT	____Screwdrivers
_____Half-Mask Respirator (Brand "A")	____Cable Cutters
_____Half-Mask Respirator (Brand "B")	____Electric Staple Gun
_____HEPA Respirator Cartridges (Brand "A")	____Staples
_____HEPA Respirator Cartridges (Brand "B")	**VACUUMS**
_____PAPR	____HEPA Vacuums
_____Batteries	____Vacuum Attachment
_____Battery Pouches	____Hose Extension
_____Battery Chargers	____Wet/Dry Vacuum
_____Supplied Air Mask	____Vacuum Filters/Bags
_____Supplied Air Belts	**SAFETY EQUIPMENT** ________Supplied
Air Hoses	____First Aid Kit
_____Carbon Monoxide (CO) Monitor	____Fire Extinguishers
_____Air Board	____Safety Lines
_____Reserve Air Bottles	____Voltage Tester
_____Reserve Air Manifold	____Ground Fault Circuit Interrupters______
5/16" Air Line, 50' Lengths	**NEGATIVE PRESSURE EQUIPMENT** ¾"
Air Line Supply	____Negative Pressure Exhaust Units
_____Moisture Traps	____Filters;Prefliters, Secondary, HEPA
_____Crowfeet Connections	____Extension Cords
_____Quick Connect Fittings (extra)	____Flexible Duct
_____Airline Manifolds	____GFCI's
_____Grade D Air Supply Compressor	
HAND TOOLS	
_____Vice Grips	
_____Pliers	
_____Crowbars	
_____Tie Wire	
_____Hammers	
_____Electric Staple Gun	
_____Staples	
_____Sweep Brooms	
_____Dust Pans, Plastic	
_____Mops	
_____String Mops	
_____Mop Bucket	

NOTE: This is not a complete check- list. Items vary from project to project. You are encouraged to prepare your own check-list.

Objectives: To become familiar with the procedures used and materials necessary for the glovebag technique of asbestos removal.

Learning Tasks: Information in this chapter should enable participants to:

- Understand the concept of localized glovebag removal.
- Become familiar with the necessary materials to perform the job.
- Recognize the importance of proper pipe lagging preparation.
- Understand the basic procedures and sequence for glovebag operations.
- Be aware of necessary precautions, work practices and personal protective equipment.
- Perform proper clean-up and disposal of asbestos-containing waste generated by this technique.

NOTE: (Please see local and state requirements for specific glovebag procedures.)

OVERVIEW OF GLOVEBAGS AND THE GLOVEBAG PROCEDURE

A glovebag typically consists of a six to eight mil polyethylene or 10 mil polyvinyl chloride (PVC) bag fitted with long sleeve gloves, a tool pouch, and a small opening which may be used for insertion of water sprayers and/or HEPA vacuum nozzles. Although glovebags can be fabricated by the user individually before each project, most contractors prefer to purchase ready-made bags.

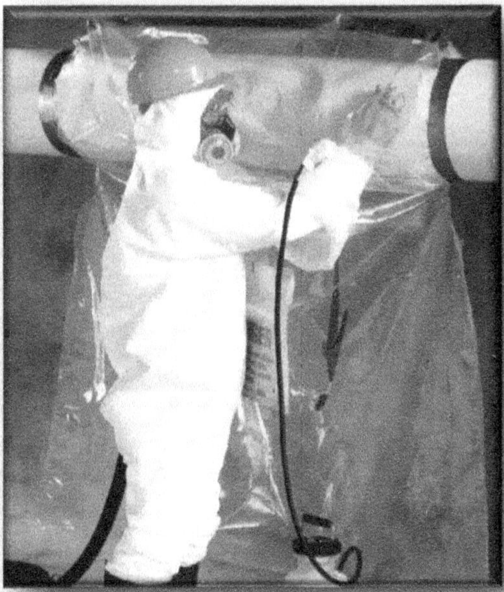

Quality, cost, size, and style of glovebags may vary depending on the manufacturer. Many manufacturers can custom design glovebags upon the request of the user. Additionally, multiple sets of sleeves can be built into one bag if the job requires it. The overall strength of the glovebag is generally determined by its thickness. Depending on the size and type of glovebag, the costs may vary significantly.

A typical glovebag is approximately forty to fifty inches wide by forty-eight to sixty inches long. Many glovebags are pre-labeled with appropriate OSHA and EPA precautions, ("Danger") in order to ensure compliance with the federal regulations. In most cases, it will be necessary to "double bag" asbestos waste. When disposing of glovebagged waste, it is necessary to ensure that the outermost bag contains the required working.

MATERIALS

In addition to the glovebag itself, there are a variety of other tools typically considered to be standard materials necessary to successfully complete the procedure. Following is a list of these "necessary" materials:

1. Glovebag (one or more depending on project size) polyethylene or PVC
2. Pump-up garden sprayer (2-3 gallon size)
3. Amended water (surfactant)
4. Duct tape (2- or 3-inch width)
5. Polyethylene disposal bags (6 mil) labeled properly per OSHA, EPA and DOT regulations
6. Smoke tubes with aspirator bulb
7. HEPA-filtered vacuum cleaner
8. Wire saw/"Flexi-saw"
9. Utility knife with retractable blade
10. Wire cutters
11. Tin snips (if aluminum jacket is present)
12. Polyethylene plastic (roll of 4 or 6 mil)
13. Dual cartridge respirators with high efficiency cartridges
14. Disposable full-body suits with good and feet covering
15. Small scrub brush (nylon brush) or scrub pad
16. Several rags
17. Wetable cloth, spray encapsulant, or other patching material
18. Asbestos danger signs and labels
19. Re-insulation materials as necessary

Look for loose material and secure/repair it prior to performing the glove-bag procedure.

Typical Glove Bag Project

Two persons should perform a safe and efficient glovebag removal project. A third person is often available to assist with supplies, keep unwanted visitors out of the area, and possibly to conduct the air monitoring. Each of these team members should have received training on the use and limitations of glovebag removal projects. They should also be included in the respiratory protection program and medical surveillance programs.

Before any work begins, all necessary materials and supplies should be brought into the work area. This work area should be roped off and danger signs posted on the perimeter to minimize the chance of visitors entering this area. Barrier tape with a preprinted asbestos warning works well for this purpose. The HVAC system serving the work area should be shut down, if possible. Employees should be trained in emergency procedures in the event the glovebag ruptures.

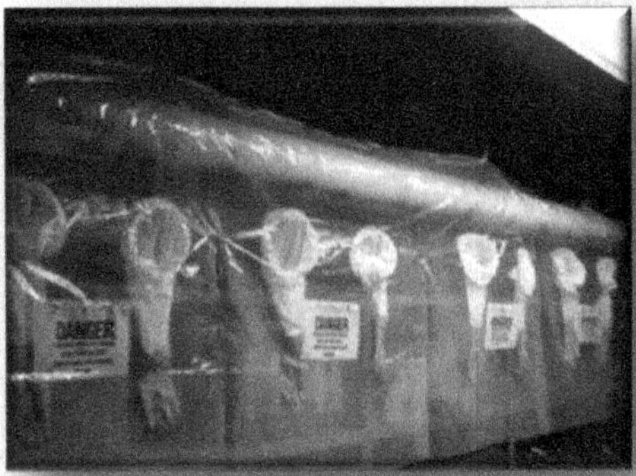

These emergency procedures usually include wet cleaning and/or HEPA vacuuming procedures, with a shower available at a remote location. With this phase completed, the following generic guidelines may be used for most pipe lagging projects. It is important to know the temperature of the pipe on which work will be conducted. If the pipe system cannot be shut down and with a shower available at a cooled before glove-bagging is done, extreme care must be taken when working with a glovebag. **OSHA does not allow the use of glovebags on pipes where the surface temperature is greater than 150°F.** Some manufacturers are currently experimenting with the development of special "high temperature" glovebags to allow removal from hot surfaces; however, it is still necessary to exercise extreme caution when dealing with these types of situations. (Polyethylene will melt at 150°F.) High-temperature glovebags that will withstand temperatures approaching 700°F are available along with the special worker protection equipment necessary for use. The use of such specialty glovebags is an unusual and hazardous operation, and is beyond the scope of this chapter.

REMOVAL PROCEDURES

1. Following the manufacturer's directions, mix the surfactant with water in the garden sprayer.

2. Have each employee put on a high efficiency cartridge respirator approved for use against asbestos and check the face-fit.

3. Have each employee put on a disposable full-body suit. Remember, the hood goes **over** the respirator straps.

4. Check the pipe on which the work will be performed and the surrounding work area. If debris is noted on the floor around the pipe, this material should be cleaned up using a HEPA vacuum and wet methods. After cleaning the area, place a poly drop cloth on the floor to catch any debris that may fall. If the pipe is damaged (broken lagging, hanging material, etc.), repair or wrap the entire length of the pipe in polyethylene sheeting and "candystripe" it with duct tape. A common error when performing glovebag work is forgetting that loose pipe lagging several feet or even several yards away from the glovebag work may be jarred loose by the removal activity. This is one of the common causes of high airborne fiber concentrations during glovebag work. If the pipe is undamaged it should be cleaned of any visible dust or debris that may be on the jacketing. It is necessary to place one layer of duct tape around the pipe at each location where the glovebag will be attached. This serves two purposes. First, it gives a good surface on which to seal the ends of the glovebag. Second, it minimizes the chance of releasing fibers when the tape at the ends of the glovebag is peeled off at the completion of the job.

5. Open the top of the glovebag and cut down the sides to accommodate the size of the pipe (about two inches longer than the pipe diameter). Some bags have zippers or two-sided tape on top and straps at each end facilitating easier installation of the bag on the pipe.

6. Place the necessary tools into the pouch located inside the glovebag. This will usually include the flexi-saw, utility knife, rags, scrub brush, wire cutters, tin snips, and spray encapsulant.

7. Place the glovebag around the chapter of pipe to be worked on. If the bag has double sided tape at the top, a preliminary seal can be made at this point. Next, fold the flap back and tape it down with a strip of duct tape. This should provide an adequate seal along the top. Next, duct tape the ends of the glovebag to

the pipe itself, at the attachment areas previously covered with plastic or duct tape.

8. Using the smoke tube and aspirator bulb, place the tube into the water porthole (two-inch opening to glovebag). By squeezing the bulb, fill the bag with visible smoke. Remove the smoke tube and twist or tape the water porthole closed. Gently squeeze the glovebag and look for smoke leaking out, especially at the top and ends of the glovebag. If leaks are found, they should be taped closed using duct tape and the bag should be re-tested with smoke.

9. Insert the wand from the water sprayer through the water. Using duct tape, tape the water porthole tightly around the wand to prevent air leakage.

10. One person places his hands into the long-sleeved gloves while the second person directs the water spray at the work and the sides of the glovebag.

11. If the chapter of pipe is covered with an aluminum jacket, this is removed first using the wire cutters to cut any bands and the tin snips to remove the aluminum. It is important to fold the sharp edges in to prevent cutting the bag when it is placed in the bottom. **Use caution to prevent cuts – these edges are sharp!**

12. With the insulation exposed, use the flexi-saw to cut the insulation at each end of the chapter to be removed inside the glovebag. (Note: A flexi-saw is a serrated heavy-gauge wire with ring-type handles at each end.) Throughout this process, water is sprayed on the cutting area to keep dust to a minimum.

13. Once the ends are cut, the chapter of insulation should be slit from end to end using the utility knife. The cut should be made along the bottom of the pipe to prevent the insulation from falling into the glovebag. Water must be continuously supplied. Again, care should be taken when using the knife not to puncture the bag. Some insulation may have wire to be clipped as well.

14. Spray all tools with water inside the bag and place back into pouch.

15. The insulation can now be lifted off the pipe. The newly exposed surfaces should be thoroughly wetted, and the insulation gently placed in the bottom of the bag. Saturate any exposed ends of pipe insulation.

16. Using the scrub brush, rags and water, scrub and wipe down the exposed pipe inside the glovebag. (Note: Inexpensive horse rub-down mittens work well for this.)

17. Spray the entire work area inside the glovebag, washing all debris down to the bottom of the bag. Use the spray encapsulant to encapsulate the exposed ends of insulation and the pipe. Note: If the pipe is a high temperature pipe, use a high temperature spray encapsulant.

18. From outside the bag, pull the tools through the glove and away from the bag and twist it to separate it from the bag. Place duct tape over the twisted portion and then cut the glove from the glovebag, cutting through the twisted/taped chapter. In this manner, the contaminated tools may be placed directly into the next glovebag without cleaning. Alternatively, the glove with the tools can be placed in a bucket of water, opened underwater, and the tools cleaned and dried without releasing asbestos into the air. (Note: Rags and the scrub brush cannot be cleaned in this manner and should be discarded with the asbestos waste.)

19. Remove the water wand from the water porthole and attach the small nozzle from the HEPA-filtered vacuum. Turn on the vacuum only briefly to collapse the bag.

20. Remove the vacuum nozzle and twist the water porthole closed and seal with duct tape.

21. With the removed insulation in the bottom of the bag, twist the bag several times and tape it to keep the material in the bottom during removal of the glovebag from the pipe. (A HEPA vacuum may help suction air out of the glovebag.)

22. Slip a 6 mil disposal bag over the glovebag (still attached to the pipe). Remove the tape and open the top of the glovebag and fold it down into the disposal bag.

23. Twist the top of the disposal bag closed, fold this over ("gooseneck" style), and seal with duct tape. Ensure that the outermost bag is appropriately labeled per OSHA, EPA, and DOT regulations.

24. Using a piece of wetable cloth which has been cut into a donut shape with the inner diameter ½ inch smaller than the diameter of the pipe, wet the cloth, and place over the exposed ends of the insulation remaining on the pipe. Wetable cloth is a plaster impregnated fiberglass webbing available at many hardware and/or plumbing supply stores.

25. Place the poly that was on the floor into another properly labeled disposal bag.

26. Remove the disposable suits and place these into the disposal bag with the poly and glove that contained the tools.

27. Using a clean damp rag, wipe the exterior of the respirator and leave the work area. Remove the respirator.

28. Asbestos-containing material must be disposed of at an approved landfill in accordance with EPA regulations.

29. Air sampling should be conducted during and after completion of glovebag projects to determine if undetected leakage occurred. Sampling should be done by qualified persons. Once a good visual inspection has been conducted, it will be possible for re-entry by unprotected personnel. Reinsulation may also occur at this point. For further information concerning sampling procedures and clearance criteria, see the chapter entitled, "Air Sampling Requirements."

OTHER SYSTEMS

Under the revised OSHA asbestos standards for the construction industry, OSHA also identifies other systems that are permitted for Class I activities. These alternative systems include negative pressure glovebox systems, a water spray process system, and the use of mini- enclosures. Requirements for the use of each of these methods are provided in the regulations.

Objectives: To understand the proper techniques for preparing the work area and setting up a decontamination unit before abatement activity begins.

Learning Tasks: Information in this chapter should enable participants to:

- Understand objectives of work area preparation.
- Become familiar with the sequence and methods for accomplishing tasks in work area preparation
- Know the functions of a decontamination unit.
- Become familiar with the basic construction of a decontamination unit.
- Know procedures for entering and leaving the work area using the decontamination unit.
- Become familiar with the necessary materials and equipment usedfor prepping the work area and building a decontamination unit.

OSHA	**29 CFR 1926.1101** Jobs involving ACM will either be: **Class I** – Removal of TSI & surfacing **Class II** – Removal of all materials except TSI & Surfacing **Class III** – Any repairs up to 1 glovebag or disposal bag **Class IV** – Cleanup where ACM is contacted but not disturbed **Unclassified operations** – any other type of activity covered under the standard's scope
EPA United States Environmental Protecti Agency	**NESHAP 40 CFR 61** Each job will be one of the following: **Demolition** – removing at least 1 load-bearing member **Renovation** – altering facility w/o removing at least 1 load –bearing member Materials involved in the job will be either: **Category I NF** – resilient flooring, asphalt roofing, gasket or packing's **Category II NF** – any non-friable not resilient flooring, asphalt roofing, gasket or packing's **RACM** – basically any currently or will become friable materials.
EPA United States Environmental Agency	**AHERA – 40 CFR 763** Include all friable asbestos abatement projects conducted in a school building or on exterior mechanical components.
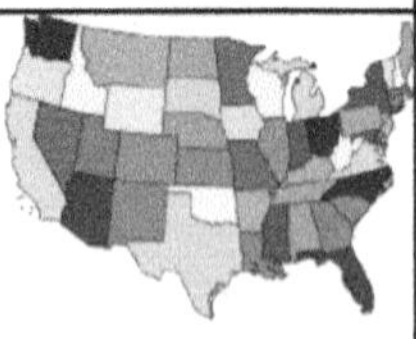	**States** Individual state regulations vary and can be more stringent than Federal requirements. State Regulations may have different trigger numbers regarding quantities and types of asbestos activities.

PREPARING THE WORK AREA

The main purpose of properly preparing the work area (where an asbestos response action is to take place) prior to a response action is to prevent exposure to airborne concentrations of asbestos fibers of both workers and building occupants. Airborne fibers that are generated by disturbance of asbestos-containing material may remain suspended in the air for long periods of time because of their small size and aerodynamic properties. These airborne asbestos fibers can migrate via air currents to other parts of the building. Preparation of the work area before an asbestos abatement project begins serves the primary purpose of containing fibers that are released within the work area. Good preparation techniques serve to protect interior finishes such as hardwood floors or carpets from water damage and to reduce cleanup efforts.

Preventing injury through appropriate safety practices is another major consideration in work area preparation (See chapter on Other Safety and Health Considerations).

Each project has unique requirements for effective preparation. For instance, the sequence of steps would probably be different for preparing a boiler room than preparing an area with asbestos material above a suspended ceiling. This may be attributed to the age of the building, the physical condition of materials, as well as HVAC system involvement. The following are general guidelines that can be modified to address specific problems encountered on an asbestos abatement project.

STEP 1 – Conduct Walkthrough Survey of the Work Area

The contractor, building owner, and project designer should make a walkthrough survey of the building or facility to inventory the ACM, note any special conditions, and photograph any existing damages. Information gained can be used to prepare an abatement plan and to aid in preparation of a realistic quotation. Complete documentation of existing conditions through the use of field notes, photographs, and videotape may benefit those involved if litigation should occur at a later date.

STEP 2 – Post Warning Signs

Warning signs that demarcate regulated work areas should be displayed at each location (entrances and exits) where airborne concentrations of asbestos may be in excess of the 0.1 f/cc permissible exposure limit or the 1.0 f/cc excursion limit. Signs should be positioned such that any person would notice the warning before entering the area and be able to take the proper necessary protective actions.

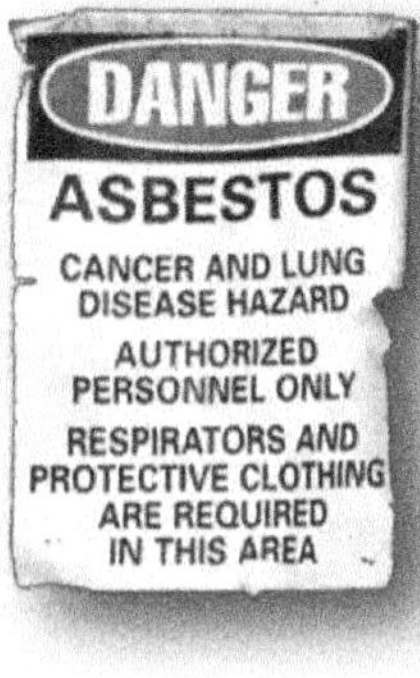

The warning signs are required to contain the following information:

(1) that asbestos is a dangerous cancer and lung disease hazard, (2) that authorized personnel only are allowed in the work area, and (3) that respirators and protective clothing are required before entering the area. See 29CFR 1926.1101(k) of the OSHA Construction Industry Standard for sign specifications. These signs are available from most safety supply companies or asbestos abatement contractor suppliers.

STEP 3 – Shut Down the Heating, Ventilating, and Air Conditioning System - HVAC

The HVAC system supplying the work area should be shut down and isolated to prevent entrainment of asbestos dust throughout the building. To avoid inadvertent activation of the HVAC system while removal operations are in progress, the control panel should be tagged and locked. Personnel need to be warned not to activate any control panels. HVAC system balancing needs to be considered to avoid over-pressurization in the occupied portions of buildings.

All vents and air ducts inside the work area should be covered and sealed with two layers of six mil polyethylene and duct tape. The first layer of polyethylene (poly) should be left in place until the area has passed final visual inspection and clearance air monitoring.

HVAC filters which may be contaminated with asbestos dust should be removed and disposed of in the same manner as the other asbestos-containing materials (see Disposal of Waste). If the filters are contaminated, the inside walls of the air ducts are probably also contaminated and the contractor should make efforts to clean or dispose of them.

STEP 4 – Clean/Remove Non-Stationary Items from the Work Area

Preparation for constructing negative-pressure enclosures, as required per 29 CFR 1926.1101(g), should begin with the cleaning of all objects in the work area. The objects should first be vacuumed with a HEPA vacuum and cleaned with amended water, unless they are made of material that will be damaged by the wetting agent. Wiping with plain water is recommended in those cases where amended water will damage the object. Non-stationary items should be removed from the work area (e.g., desks, chairs, rugs, and light fixtures) to ensure that these objects do not become contaminated with asbestos. Drapes should be removed for cleaning or disposal. Carpets contaminated with debris or suspected of being contaminated should be disposed of as asbestos-containing waste.

Workers involved with the cleaning and removal should at least wear a half mask HEPA filter dual cartridge respirator and disposable clothing when initial preparation work is being carried out.

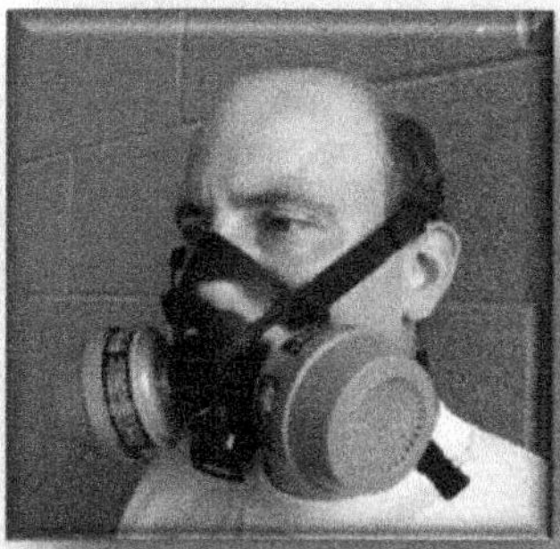

STEP 5 – Cover and Seal Stationary Items with Polyethylene

Before the asbestos response action begins, objects that cannot be removed from the asbestos-contaminated work area must be HEPA vacuumed or wet-wiped and covered with 6 mil-thick poly* sheeting. They should be securely tape with duct tape or plastic tape to achieve an air tight seal around the object and to ensure that they do not become contaminated during the removal project. Items not being removed may include large pieces of machinery, blackboards, water fountains, toilets, etc. Use of two layers of poly* is a good recommended practice.

A specific outline of items to be covered and sealed is included here. *(Fire Retardant)

STEP 6 – Apply Critical Barriers

A. Windows and Doors

The edges of all the windows should be sealed with 2" or 3" wide high quality duct tape. After the edges have been taped, the windows should be covered and sealed with six mil poly* and duct tape. Covering windows and all other doors not being used during abatement with a separate layer of poly (called a critical barrier) before covering the walls provides a back-up layer of protection and saves time in installation because it reduces the number of edges of poly that must be cut and taped. A single entrance to be used for access and egress to the work area should be selected. This would most likely be the decontamination area that is discussed later in this chapter.

B. Floor

Six mil poly* sheets should be used to cover the floor in the work area. Several sheets will need to be seamed together with spray adhesive and duct tape. To check the integrity of the seal, blue or red carpenter's chalk may be placed beneath the seam line. If a water leak occurs, the seam line will darken in color. Any leaks that occur should be promptly cleaned up. The poly floor sheets would be cut and peeled back to access the wet area. After mopping up the water and any contamination that leaked through, the area should be wet-wiped with clean rags. After the area dries, it is HEPA-vacuumed, and the peeled-back sheets are put back in place and sealed with duct tape. An additional "patch" sheet can be placed over this area and sealed with tape to provide extra protection.

After joining the sheets of poly together, the floor covering should be cut to the proper dimensions, allowing the poly to extend twenty-four inches up the wall, all the way around the room. The poly should be flush with the walls at each corner to prevent damage by foot traffic.

When the first layer of poly has been secured in place, the walls are covered with poly and a second layer should be laid on the floor with the seams of the first and second layers offset. The second layer of poly should extend a few inches above the first layer on the wall and be secured with duct tape.

Potential slippery spots may be encountered when covering stairs or ramps and care must be taken to provide traction for foot traffic. Wet poly is very slippery and can create serious tripping hazards. To provide better footing, masking tape or thin wood strips can be placed on top of the poly to provide rough surfaces in these areas.

C. Walls

After the first layer of polyethylene has covered the floors and stationary objects, multiple layers of four mil polyethylene are used to cover the walls. The lighter-weight four mil is easier to hang and keep in place than the heavier six mil.

The sheets of four mil poly should be hung from the top of the wall a few inches below the ceiling and should extend across the floor area until they meet in the center of the area, where they are taped to form a single layer of material encasing the entire room, except for the ceiling. Overlapping of the vertical sheets will be necessary; the seams should be sealed with adhesive duct tape.

Duct tape alone will not support the weight of the poly after exposure to the varying environmental conditions that occur inside the work area. The sheets may be hung using a combination of nails and furring strips (small wood blocks), or adhesive and staples, and sealed with four-inch duct tape. Nails may cause some minor damage to the interior finish; however, it is usually more time efficient to touch up the nail holes than to repeatedly repair fallen barriers.

D. Light Fixtures

Light fixtures may have to be removed or detached and suspended (baling wire works well) to gain access to asbestos- containing material.

Before beginning this task, the electrical supply should be shut off, locked, and tagged. The light fixtures should be wet-wiped before they are removed from the area. If it is not feasible to remove the fixtures, they should be wet-wiped and completely enclosed with poly.

STEP 7 – Locate and Secure the Electrical System

Amended water is typically used to saturate asbestos-containing sprayed-on material prior to removal. This creates a humid environment with damp to very wet and slippery floors. To eliminate the potential for a shock hazard, the electrical supply to the work area should be de-energized, locked out, and tagged before removal operations begin. The following items need to be addressed before removal actually begins:

1. Identify and de-energize electrical circuits in the work area.

2. Lock the breaker box after the system has been shut down and place a warning tag on the box. The breaker box can't be locked if it contains energized circuits for non-work areas; individual breakers may have to be locked out. Custodial personnel should be consulted about electrical distribution to other areas of an occupied building.

3. Make provisions for supplying the work area with electricity from outside the work area, which is equipped with a ground-fault-interrupt system.

4. If the electrical supply cannot be disconnected, energized parts must be insulated or guarded from employee contact and any other conductive object.

STEP 8 – Securing the Work Area

The work area should be secured to prevent contamination from spreading beyond the work area. All entrances should be secured when removal operations are not in progress. Provisions must also be made to secure the decontamination station entrance when no one is on the job site. Security guards may be a reasonable precaution, depending on the nature of the project. When the work area is occupied, padlocks must be removed to permit emergency escape routes. Arrows should be taped or painted on the poly covered walls to indicate the location of exits.

Nonessential personnel should not be permitted to enter the work area. A job log should be maintained onsite (in the clean area) for recording who enters the work area and the time each person enters and exits the work zone. The project supervisor (or designee) should be sure that the log is maintained on a daily basis.

Step 9 – Establishing a Decontamination Unit

Employers involved in asbestos removal, demolition, or renovation operations must provide their employees with hygiene facilities to be used to decontaminate asbestos-exposed workers, equipment, and clothing before such employees leave the work area. The decontamination station is designed to allow passage to and from the work area during asbestos operations with minimal leakage of asbestos-containing dust to the outside. A typical decontamination unit consists of a clean change room, a shower, and an equipment room separated by airlocks. The work area will be kept under negative air pressure 24 hours a day, including weekends, until final air clearance is achieved.

Materials used to construct a typical unit include: 2-inch by 4-inch lumber for the frame, ¼ inch to ½ inch plywood or six mil poly for the walls, duct tape, staples, and nails. The floor should be covered with three layers of six mil polyethylene.

Chapters of the decontamination unit can be built separately to allow for easy disassembly and re-use (frames only, not poly) at other areas of buildings or at other job sites. Designs of decontamination stations may vary with each project depending on the size of the crew and the physical constraints imposed by the facility.

Customized trailers that can be readily moved from one location to the next are also used as decontamination stations. These units typically cost $20,000- $50,000 depending on the size and features. A company conducting work at many different locations would probably recover this initial investment over time.

Whether a decontamination station is constructed on site or is in the form of a trailer, the basic design components are the same.

Clean Room

As described in the OSHA Asbestos Standard for the Construction Industry (1926.1101), the clean room is an uncontaminated room having facilities for the storage of employees' street clothing and uncontaminated materials and equipment. It is an area in which employees remove their street clothes, store them, and don their respirators and disposable protective clothing. This room is where workers dress in clean clothes after showering. Furnishings for the clean room should include: benches, lockers for clothes and valuables, and nails or hooks for hanging respirators.

Extra disposable coveralls and towels can be stored in the clean change room.

Shower Room

The shower should have on either side of it, two airlocks, with both the clean and dirty change rooms on either side of the airlocks. Workers pass through the shower room on their way to the removal area and use the showers on their way out after leaving contaminated clothing in the equipment room. Although most job specifications require only a single shower head, installation of multiple showers may be time and cost effective if the work crew is large. Cold and hot water are required by OSHA with separate controls.

Shower wastewater should be drained, collected, and filtered through a system before disposal into the sanitary sewer. A system containing a series of several filters with progressively smaller pore size (100, 50, 5 micron) is recommended to avoid rapid initial clogging of filtration system by larger particles. Wastewater may need to be retained in sealed barrels or containers and/or holding tanks for appropriate disposal. Some states have written specifications for handling shower wastewater. Ask you instructor on the state regulations.

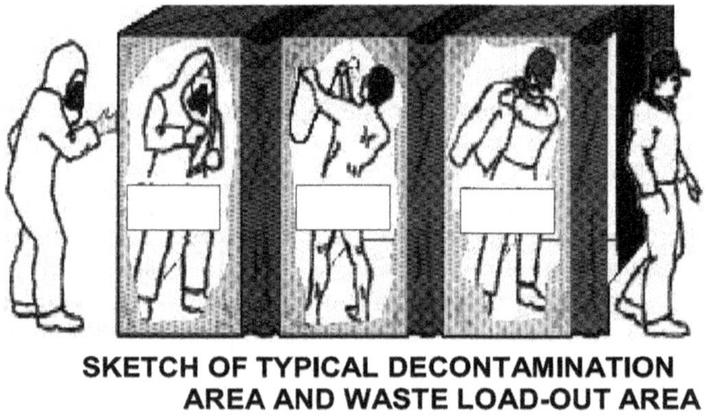

SKETCH OF TYPICAL DECONTAMINATION AREA AND WASTE LOAD-OUT AREA

please label each room above

SEQUENCE OF PROCEDURES FOR ENTERING and EXITING the WORK AREA

IN THE CLEAN ROOM, WORKER:

1. Enters clean room
2. Removes clothing, places in locker
3. Puts on nylon swim suit (optional)
4. Puts on clean coveralls
5. If separate disposable foot coverings are used, these are put on
6. Applies tape around ankles, wrists, etc.
7. Inspects respirator, puts it on, checks fit
8. Puts on hood over respirator headstraps
9. Proceeds to equipment room

IN THE EQUIPMENT ROOM, WORKER:

10. Puts on any additional clothing – deck shoes, hard hat, etc.
11. Collects necessary tools and proceeds to WORK AREA

IN THE WORK AREA, WORKER:

12. HEPA Vacuums off contamination

IN THE EQUIPMENT ROOM, WORKER:

13. Removes all clothing except respirator
14. Places disposable protective clothing in a bag or bin
15. Stores any other contaminated articles
16. Proceeds to shower

IN THE SHOWER, WORKER:

17. Washes respirator and soaks filters (without removing)
18. Removes respirator, washes with soap and water, disposes of contaminated filters in receptacle in shower or reaches into dirty room to place filters into easily accessible disposal bag
19. Washes swim suit
20. Thoroughly washes body and hair

IN THE CLEAN ROOM, WORKER:

21. Dries off, dresses in clean coveralls or street clothes
22. Cleans and dries respirator, replaces filters (if applicable)

Equipment Room

This area, also called the dirty change room, is the contaminated area where workers remove their protective coveralls and where equipment, boots or shoes, hardhats, goggles, and any additional contaminated work clothes are stored. Workers place disposable clothing such as coveralls, booties and hoods in bins before leaving this area for the shower room. Respirators are worn into the shower and thoroughly soaked with water before they are taken off. The equipment room will probably require cleanup several times daily to prevent asbestos materials from being tracked into the shower and clean rooms.

Airlocks

Airlocks are formed by overlapping three sheets of polyethylene at the exit of one room and two sheets at the entrance to the next room with two feet of space between the barriers. There are various methods used for constructing airlocks including a hatch type construction and a slit and cover design.

Step 10 – Establishing a Waste Load-Out Area

The waste load-out area (separate from the decontamination unit and not used for personnel egress) is used as a short-term storage area for bagged waste and as a port for transferring waste to the truck. An enclosure can be constructed to form an airlock between the exit of the load-out area and an enclosed truck.

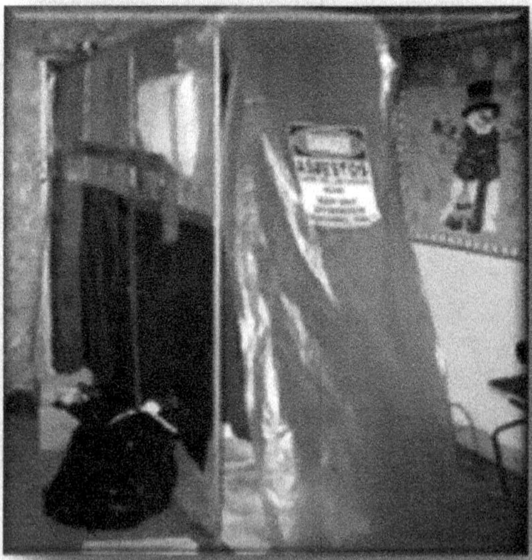

The outside of the waste containers should be free of all contaminated material before removal from the work area. Gross contamination should be wiped or scraped off containers before they are placed in the load-out area. Any remaining contamination should be removed by wet-wiping; the bagged material can be placed in a second clean bag.

The clean room, shower, and equipment room must be sealed completely to ensure that the sole source of air flow through these areas originates from uncontaminated areas outside the asbestos removal, demolition, or renovation enclosure. After construction of the enclosure is completed, a ventilation system(s) should be installed to create a negative pressure within the enclosure with respect to the area outside the enclosure. Such ventilation systems are discussed in detail in the Chapter "Confining and Minimizing Airborne Fibers."

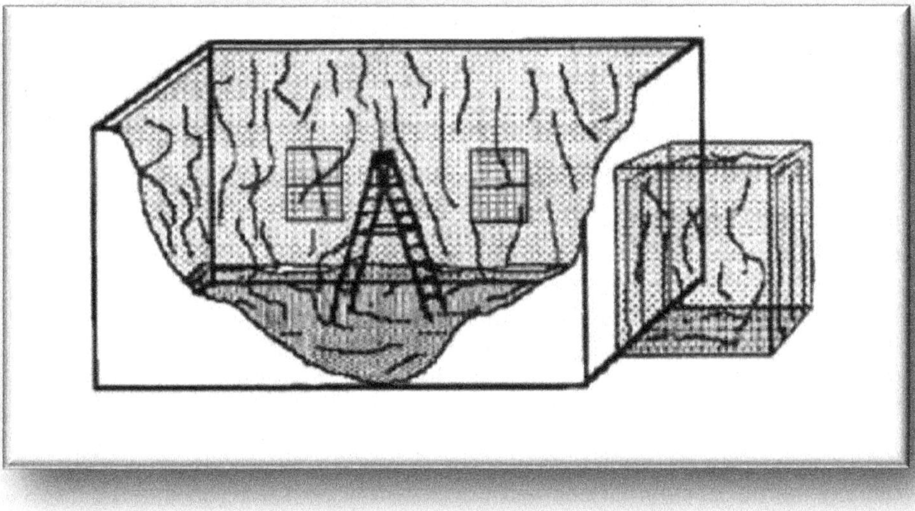

Mini-Enclosure

A mini-enclosure is usually a polyethylene enclosure around a work area. Mini-enclosures are sealed enclosures used to protect the facility environment as a secondary means to help, or attempt to, contain fibers or debris generated during the work.

Mini-enclosures also serve to provide a visual barrier between the workers and any other personnel around the work area. As noted above, careful work practices should be the primary means of fiber control during the work in order to prevent gross contamination of the mini-enclosure.

It is sometimes appropriate to extend mini-enclosures above ceilings, such as by using polyethylene sheet and framing taped together to provide enclosure around the work area. The mini-enclosure should not contact ACM covered surfaces. The construction will vary depending on whether the enclosure will be attached to pipes, conduit, metal hangers, or some other form of existing construction.

There are a variety of commercially available types of mini-enclosures, including prefabricated pop-up boxes and adjustable framework assemblies to permit different sizes of enclosures to be constructed.

Disposable liners for mini-enclosures (to facilitate set up and dismantling of the enclosure) are available from some manufacturers. It might be beneficial to construct or purchase a portable mini-enclosure unit that works for the typical conditions found in a given facility.

It is recommended that two workers be used to set up and operate mini-enclosures. To construct a mini-enclosure, erect a framework of wood, PVC piping or metal framing that will enclose the work area and be large enough for one person to work inside. The minimum width and depth of the enclosure should be at least 3 feet (1 meter). The height of the enclosure will vary depending upon the work to be performed and the height of the work area. A larger enclosure is preferable where space permits. However, if the enclosure is too large, the final cleaning process will require more time. A mini-enclosure can include a separate 3 foot by 3 foot by 7 foot (1 x 1 x 2.1 meters) change room, with curtain doorways, attached to the mini-enclosure for changing and removing protective clothing.

If an entire room will be enclosed for performing work, the framework is usually not necessary, unless wall surfaces will be damaged by tape used to support polyethylene. A room can be enclosed for O & M work by installing one layer of polyethylene sheet on the walls and floor of the room.

If the work to be performed is in an elevated location, the enclosure (and change room, if used) should be erected on a scaffold platform large enough to support the enclosure, change room (if used), and a step off area outside the enclosure.

Refer to OSHA regulations 29 CFR 1910.28 and 29 CFR 1926.451 concerning scaffold requirements. Any ladders and/or scaffolds used must be built and used in conformance with the OSHA construction standards, and applicable state and local standards.

Cover the floor and the framework for the enclosure and change room with one layer of polyethylene attached using duct tape. A second layer of polyethylene laid on the floor might facilitate clean up work, or reduce the possibility of tearing the polyethylene if equipment is used (do not use two layers under the legs of ladders). Construct curtain doorways between the change room and the enclosure and between the change room and the area outside the change room. A curtain doorway is made of three overlapping sheets of polyethylene. Attach sheets to framework at top and one side. The middle sheet should be attached on one side, and the inner and outer sheets attached on the other side. A sheet of polyethylene approximately 5 feet by 5 feet (1.5 meters by 1.5 meters) or larger should be installed outside the change room for use as a step off area and as a place to put decontaminated materials removed from the work area.

Mini-enclosures should be constructed with a ceiling of polyethylene if work will not be performed above the enclosure. If work is to be performed above the enclosure and the ceiling is not ACM, the enclosure should extend to and be sealed to the ceiling or grid system. If the enclosure is below an ACM finished surface, use one of the following methods:

1. If ACM cannot be contacted, the enclosure should be separated from the ceiling by a narrow space.

2. If ACM will withstand contact without damage and is in good condition, foam tape (1" (25mm) or thicker) can be placed on the top edge of the enclosure. Gently lift enclosure into place until sufficient contact is made to provide a seal to the surface.

After enclosure is in place, check for, and clean up any debris generated by enclosure installation.

Mini-enclosures should be set up with a negative pressure system as described to reduce the possibility of fibers being released from the enclosure and to filter the air inside the enclosure.

Objectives: To provide instruction to participants on the most effective methods for containment of asbestos fibers during an asbestos abatement project.

Learning Tasks: Information in this section should enable participants to:

- Understand the primary methods used to contain and minimize airborne fiber concentrations
- Contain and minimize airborne fiber concentrations during an asbestos abatement project.
- Know principles and procedures for setting up a negative air filtration system on an abatement project.
- Become familiar with the use and limitations of negative air filtration units.
- Understand the application and use of wet removal techniques.
- Become familiar with proper procedures and equipment for removal of asbestos-containing sprayed and troweled-on friable insulation material.
- Become familiar with proper procedures and equipment for removal of asbestos-containing insulation from pipes, tanks, and boilers.

CONFINING AND MINIMIZING AIRBORNE FIBERS

The preparation phase of an asbestos abatement project is directed toward containing the airborne fibers that will be generated during removal, primarily by constructing barriers with polyethylene sheeting. This containment effort, along with measures to minimize airborne fiber concentrations, is continued throughout the removal phase. The primary methods for contaminant control are the use of wet removal techniques and the use of negative pressure filtration systems accompanied by continuous cleanup in a work area sealed with polyethylene.

NEGATIVE PRESSURE FILTRATION SYSTEMS

Figure - XI-1

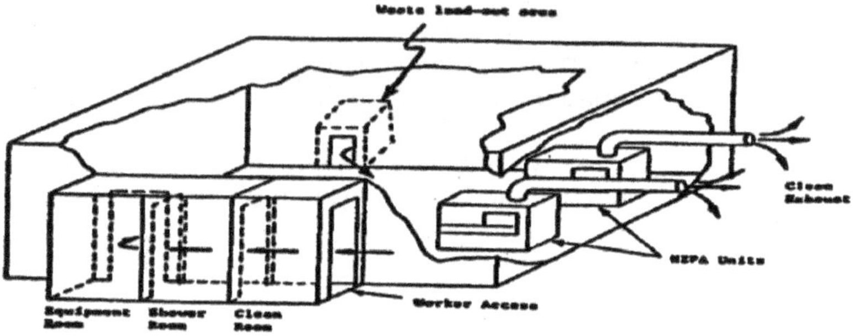

Note: Please see state regulations diagrams for negative pressure filtration system. (See instructor)

Figure XI-2

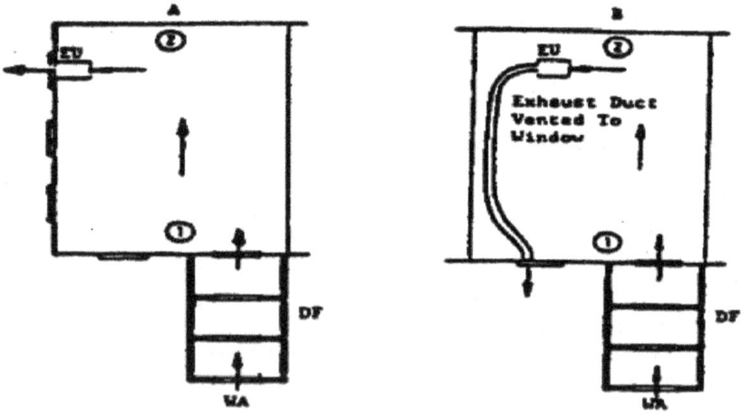

The planning strategy for the use of negative pressure systems in abatement work includes two main goals.

1. Changing air within the containment area at a minimum of every 15 minutes while filtering the exhausted air through high efficiency particulate air (HEPA) filters.

2. Establishing conditions in which air from all portions of the sealed zone is being pulled toward the negative pressure fans and HEPA filters.

Negative pressure systems should be used on an abatement project to accomplish several positive effects.

1. Containment of airborne fibers even if the barrier is ripped or punctured.

2. Lower concentration of airborne fibers in the work area.

3. Worker comfort and increased productivity.

4. Improved efficiency in final cleanup.

Negative pressure filtration units are known by several different names including Micro-Trap™, Red Baron™, Hog™, micro-filter, HEPA units, and negative pressure system. Prototypes for use in asbestos abatement were developed in the latter 1970s. The concept of air filtration systems as a primary control technique on asbestos abatement projects was adopted by EPA. A general discussion on negative air systems is provided in the following pages.

RECOMMENDED SPECIFICATIONS AND OPERATING PROCEDURES FOR THE USE OF NEGATIVE PRESSURE SYSTEMS FOR ASBESTOS ABATEMENT*

This section provides guidelines for the use of negative pressure systems in removing asbestos-containing materials from buildings. The manufacturer's instructions for equipment use should be followed for negative air filtration units, as well as all other equipment discussed in this manual. A negative pressure system is one in which the static air pressure in an enclosed work area is lower than that of the environment outside the containment barriers.

The pressure gradient is maintained by moving air from the work area to the environment outside the area via powered exhaust equipment (negative air filtration unit) at a rate that will support the desired air flow and pressure differential. Thus, the air moves into the work area through designated access spaces and any other barrier openings. Exhaust air is filtered by a high- efficiency particulate air (HEPA) filter to remove asbestos fibers.

The use of negative pressure during asbestos removal helps protect against the large-scale release of fibers to the surrounding area in case of a breach in the containment barrier. A negative pressure system also can reduce the concentration of airborne asbestos in the work area by increasing the dilution ventilation rate (i.e., diluting contaminated air in the work area with uncontaminated air from outside) and exhausting contaminated air through HEPA filters. The circulation of fresh air through the work area reportedly also improves worker comfort by increasing the cooling effect, which may aid the removal process by increasing job productivity.

MATERIALS AND EQUIPMENT

The Portable, HEPA-Filtered, Powered Exhaust Unit

The exhaust unit establishes lower air pressure inside than outside the enclosed work area during asbestos abatement by moving air from the contained work area to the outside. Basically, a unit consists of a cabinet with an opening at each end, one for air intake and one for exhaust. A fan and a series of filters are arranged inside the cabinet between the openings. The fan draws contaminated air through the intake and filters and discharges clean air through the exhaust. Portable exhaust units used for negative pressure systems in asbestos abatement projects should meet the following specifications.

STRUCTURAL SPECIFICATIONS

The cabinet should be ruggedly constructed and made of durable materials to withstand damage from rough handling and transportation. The width of the cabinet should be less than 30 inches to fit through standard-size doorways. Appropriate cabinet seals should prevent asbestos-containing dust from being emitted during use, transport, or maintenance. There should be easy access to all air filters from the intake end. The filters must be easy to replace. The unit should be mounted on casters or wheels so it can be easily moved. It also should be accessible for easy cleaning.

MECHANICAL SPECIFICATIONS

Fans

The fan for each unit should be sized to draw a desired air volume through the filters in the unit at a specified static pressure drop (see manufacturer's literature for this information). The unit should have an air-handling capacity of at least 1,000 to 2,000 cubic feet per minute (CFM or ft^3/min) (under "clean" filter conditions). The fan should be of the centrifugal type.

For large-scale abatement projects, where the use of a larger capacity, specially designed exhaust system may be more practical than several smaller units, the fan should be appropriately sized according to the proper load capacity established for the application, i.e.,

$$\text{Total } ft^3/\text{min (load)} = \frac{(\text{Volume of work area in } ft^3)\ (\text{air changes/hour})}{60 \text{ min/hour}}$$

Smaller-capacity units (e.g., 500 ft^3/min) equipped with appropriately sized fans and filters may be used to ventilate smaller work areas. The desired air flow could be achieved with several units.

FILTERS

The final filter must be the HEPA type. Each filter should have a standard nominal rating of at least 1,100 ft^3/min with a maximum pressure drop of 1 inch H_2O clean resistance. This pressure drop will increase as the filters load and the manufacturer's literature will indicate a clean filter pressure drop and a recommended maximum allowable pressure drop for dirty filters. The filter media (folded into closely pleated panels) must be completely sealed on all edges with a structurally rigid frame and cross-braced as required to prevent air bypassing the filter. Exact dimensions of the filter should correspond with the dimensions of the filter housing inside the cabinet or the dimensions of the filter-holding frame. The recommended standard size HEPA filter is 24 inches high x 24 inches wide x 11½ inches deep. The overall dimensions and squareness should be within 1/8 inch.

A continuous rubber gasket must be located between the filter and the filter housing to form a tight seal. The size of the gasket material is dependent upon the manufacturer. (Some manufacturers use gaskets that are approximately ¼ inch thick and ¾ inch wide.) This gasket should be checked periodically for cracks and gaps. Any break in this gasket may permit significant leakage of contaminated air. Leaks in the gasket or filter will be indicated by lower than normal "clean resistance" pressure.

Each filter should be individually tested and certified by the manufacturer to have an efficiency of not less than 99.97 percent when challenged with 0.3 micrometers (μm) dioctylphthalate (DOP) aerosol. Testing should be in accordance with Military Standard Number 282 and Army Instruction Manual 136-300-175A.

Each filter should bear a UL586 label to indicate ability to perform under specific conditions. Each filter should be marked with: the name of the manufacturer, serial number, air flow rating, efficiency and resistance, and the direction of test air flow.

Prefilters, which protect the final filter by removing the larger particles, are recommended to prolong the operating life of the HEPA filter. Prefilters prevent the premature loading of the HEPA filter. They can also save energy and cost. One (minimum) or two (preferred) stages of prefiltration may be used. The first-stage prefilter should be a low-efficiency type (e.g., for particles 10 μm and larger). The second-stage (or intermediate) filter should have a medium efficiency (e.g., effective for particles down to 5 μm). Various types of filters and filter media for prefiltration applications are available from many manufacturers. Prefilters and intermediate filters should be installed either on or in the intake grid of the unit and held in place with special housings or clamps.

Instrumentation

Each unit should be equipped with a Magnehelic gauge or manometer to measure the pressure drop across the filters which would indicate when filters have become loaded and need to be changed. The static pressure across the filters (resistance) increases as they become loaded with dust, affecting the ability of the unit to move air at it rated capacity.

ELECTRICAL

General

The electrical system should have a remote fuse disconnect. The fan motor should be totally enclosed, fan-cooled, and the non-overloading type. The unit may use a standard 115-V, single-phase, 60-cycle service. All electrical components must be approved by the National Electrical Manufacturers Association (NEMA) and Underwriter's Laboratories (UL).

Fans

The motor, fan, fan housing, and cabinet should be grounded. All units should have an electrical (or mechanical) lockout to prevent the fan from operating without a HEPA filter.

Instrumentation

An automatic shutdown system that would stop the fan in the event of a major rupture in the HEPA filter or blocked air discharge is recommended. Optional warning lights are recommended to indicate normal operation, too high of a pressure drop across the filters (i.e., filter overloading), and too low of a pressure drop (i.e., major rupture in HEPA filter or obstructed discharge). Elapsed time meters may also be purchased to show the total accumulated hours of operation of the negative pressure units.

SETUP AND USE OF A NEGATIVE PRESSURE SYSTEM

Determining Approximate Ventilation Requirements for a Work Area

Experience with negative pressure systems on asbestos abatement projects indicates a recommended minimum rate of one air change every 15 minutes (Many states require four air changes per hour). The volume (in ft^3) of the work area is determined by multiplying the floor area by the ceiling height. The total volumetric air flow requirement (in ft^3/min) for the work area is determined by dividing this volume by the recommended air change rate (i.e., one air change every 15 minutes).*

Total ft^3/min = Volume of work area (in ft^3)/15 min**

The number of units needed for the application is determined by dividing the total ft^3/min by the rated capacity of the exhaust unit.

$$\text{Number of units needed} = \frac{\text{Total ft}^3\text{/min.}}{\text{Capacity of unit (ft}^3\text{/min.)}}$$

Location of Exhaust Units

The exhaust unit(s) should be located so that makeup air enters the work area primarily through the decontamination facility and traverses the work area as much as possible. This may be accomplished by positioning the exhaust unit(s) at a maximum distance from the worker access opening or other makeup air sources.

Wherever practical, work area exhaust units can be located on the floor in or near unused exterior doorways or windows. The end of the unit or its exhaust duct should be placed through an opening in the plastic barrier or wall covering. The plastic around the unit or duct should then be sealed with tape.

Each unit must have temporary electrical power (115V A.C.). If necessary, three- wire extension cords can supply power to a unit. The cords must be in continuous lengths (without splice), in good condition, and should not be more than 100 feet long. They must not be fastened with staples, hung from nails, or suspended by wire. Extension cords should be suspended off the floor and out of workers' way to protect the cords from traffic, sharp objects, and pinching.

Wherever possible, exhaust units should be vented to the outside of the building. This may involve the use of additional lengths of flexible or rigid duct connected to the air outlet and routed to the nearest outside opening. Windowpanes may have to be removed temporarily.

Additional makeup air may be necessary to avoid creating too high of a pressure differential, which could cause the plastic coverings and temporary barriers to detach from the walls and fall. Additional makeup air also may be needed to move air most effectively though the work area. Supplemental makeup air inlets may be made by making openings in the plastic sheeting that allow air from outside the building into the work area. Auxiliary makeup air inlets should be as far as possible from the exhaust unit(s) (e.g., on an opposite wall), off the floor (preferably near the ceiling), and away from barriers that separate the work area from occupied clean areas. They should be constructed in such a fashion (using weighted flaps, etc) that allows the openings to be sealed in case of accidental pressure differential loss. Also, the openings should be resealed whenever the negative pressure system is turned off after removal has started. Because the pressure differential (and ultimately the effectiveness of the system) is affected by the adequacy of makeup air, the number of auxiliary air inlets should be designed and placed in order to maintain adequate pressure differential and to maximize air circulation throughout the work area. Figure XI-1 presents examples of negative pressure system denoting the location of HEPA-filtered exhaust units and the direction of air flow. Figure XI-2 is a schematic representation of negative pressure systems denoting the location of HEPA- filtered exhaust units and the direction of air flow. Figure XI-2 is a schematic representation of negative air HEPA system in place.

USE OF THE NEGATIVE PRESSURE

SYSTEM Testing the System

The negative pressure system should be tested before any asbestos-containing material is wetted or removed. After the work area has been prepared, the decontamination facility set up, and the exhaust units(s) installed, the unit(s) should be started (one at a time). Observe the barriers and plastic sheeting.

The plastic curtains of the decontamination facility should move slightly in toward the work area. The use of ventilation smoke tubes and an aspirator bulb is another easy and inexpensive way to visually check system performance and direction of air flow through openings in the barrier. For example, smoke emitted on the inside of the work area at a barrier should not leak outward. Smoke emitted in the shower room of the decontamination unit should move inward to the work area. Smoke tubes can also be used to check if air flow is moving inward at high and low levels of the work area.

NEGATIVE PRESSURE SYSTEMS

DF, Decontamination Facility
EU, Exhaust unit
WA, Worker Access
A, Single-room work area with multiple windows
B, Single-room work area with single window near entrance
C, Single-room work area with exhaust unit placed on the outside of the building;
D, Large single-room work area with windows and auxiliary makeup air source
Arrows denote direction of air flow.
Circled numbers indicate progression of removal sequence.

NEGATIVE AIR HEPA SYSTEM IN PLACE

Another test method for negative pressure is to use a Magnehelic gauge (or other instrument) to measure the static pressure differential across the barrier. The measuring device must be sensitive enough to detect a relatively low pressure drop. A Magnehelic gauge with a scale of 0 to 0.25 or 0.50 inch of H_2O and 0.005 or 0.01 inch graduations is generally adequate. The pressure drop across the barrier is measured from the outside by punching a small hole in the plastic barrier and inserting one end of a piece of rubber tubing.

The other end of the tubing is connected to the "low pressure" tap of the instrument. The "high pressure" tap must be open to the atmosphere. The pressure is read directly from the scale. After the test is completed, the hole in the barrier must be patched. Instruments are also available that monitor the pressure drop continuously. These units can be connected to a strip chart recorder to provide continuous documentation of negative pressure. An audible and/or visible alarm may be used to alert the project manager of a severe drop in pressure.

Typically, a pressure drop of 0.03 inches of water is maintained throughout the asbestos abatement project (this pressure drop is affected by the air change rate). The U.S. Occupational Safety and Health Administration (OSHA) require a pressure differential of 0.02 inches of water.
(Note: Some states require the use of a manometer to test negative pressure.)

Use of System during Removal Operations

The exhaust units should be started before any asbestos-containing material is disturbed. After removal has begun, the units should run continuously to maintain a constant negative pressure until final air clearance has been achieved. The units **should not be turned off** at the end of the work shift or when removal operations temporarily stop.

Employees should start removing the asbestos material at a location farthest from the exhaust units and work toward them. If an electric power failure occurs, removal must stop immediately and should not resume until power is restored and exhaust units are operating again.

Because airborne asbestos fibers are microscopic in size and tend to remain in suspension for a long time, the exhaust units must keep operating throughout the entire abatement project, decontamination, and final clearance processes.

Leaving the negative pressure system operating during the final cleanup and clearance process allows the suspended fibers the potential to be "cleaned" from the air. Also, until the results of final clearance air samples are known, the confining and minimizing aspects of negative pressure filtration are needed to ensure leakage of contaminated air outside the enclosure does not occur.

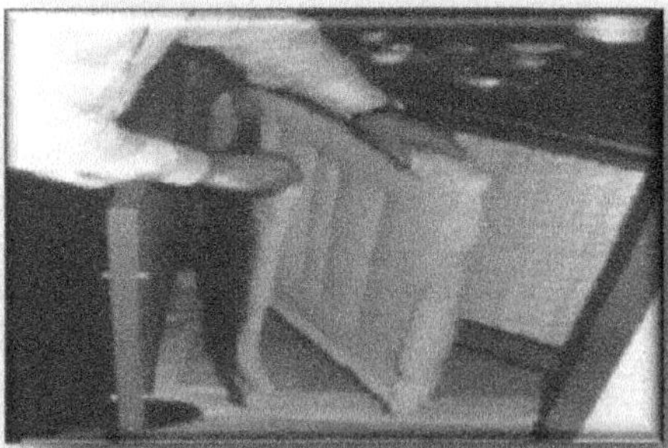

To ensure continuous operation (and therefore continuous negative pressure differential), a spare negative pressure exhaust unit(s) should be readily available at all times.

Filter Replacement

All filters must be accessible from the work area or "contaminated" side of the barrier. Thus, personnel responsible for changing filters while the negative pressure system is in use should wear approved respirators and other protective equipment. The operating life of a HEPA filter depends on the level of particulate contamination in the environment in which it is used. During use, filters will become loaded with dust, which increases resistance to air flow and diminishes the air-handling capacity of the unit. The difference in pressure drop across the filters between "clean" and "loaded" conditions is a convenient means of estimating the extent of air-flow resistance and determining when the filters should be replaced.

When the pressure drop across the filters (as determined by the Magnehelic gauge or manometer on the unit) exceeds the pressure specified by the manufacturer, the pre-filter should be replaced first. The pre-filter, which fan suction will generally hold in place on the intake grill, should be removed with the unit running by carefully rolling or folding in its sides. Any dust dislodged from the pre-filter during removal will be collected on the intermediate filter. The used pre-filter should be wetted and placed inside a six mil plastic bag, sealed and labeled, and disposed of as asbestos waste. A new pre-filter is then placed on the intake grill. Filters for pre-filtration applications may be purchased as individual precut panels or in a roll of specified width that must be cut to size.

If the pressure drop still exceeds the manufacturer's specified pressure after the pre-filter has been replaced, the intermediate filter is replaced. With the unit operating, the pre-filter should be removed, the intake grill or filter access opened, and the intermediate filter removed. Any dust dislodged from the intermediate filter during removal will be collected on the HEPA filter. The used intermediate filter should be wetted and placed in a sealable plastic bag (appropriately labeled) and disposed of as asbestos waste. The gasket between the filter and the housing should be inspected for any gaps or cracks. Worn gaskets should be replaced as needed. A new HEPA filter (structurally identical to the original filter) should then be installed. The intake grill and intermediate filter should be put back in place, the unit turned on, and the pre-filter positioned on the intake grill. Whenever the HEPA filter is replaced, the pre-filter and intermediate filter should also be replaced.

(**NOTE**: At least one backup negative air unit is required to be connected to the negative pressure enclosure.)

When several exhaust units are used to ventilate a work area, negative pressure can be maintained during the HEPA filter replacement and the direction of air flow into the work area will be maintained. If only two exhaust units are operating on-site, a backup unit should be available and operating before an original unit is shut down for HEPA filter replacement. An abatement enclosure should <u>never</u> have only one exhaust unit operating.

A failure of this sole unit for any reason would eliminate the negative pressure in the work area. Thus, the risk of asbestos fiber release to the outside environments controlled with additional unit(s).

Any filters used in the system may be replaced more frequently than the pressure drop across the filters indicates is necessary. Experience has shown that profilers, for example, should be replaced two to four times a day or when accumulations of particulate matter become visible. Intermediate filters must be replaced once every day or so, and the HEPA filter may be replaced at the beginning of each new project. (Used filters must be disposed of as asbestos- containing waste).

Conditions in the work area dictate the frequency of filter changes. In a work area where fiber release is effectively controlled by thorough wetting and good work practices, fewer filter changes may be required than in work areas where the removal process is not well controlled. It should also be noted that the collection efficiency of a filter generally improves as particulate accumulates on it. Thus, filters can be used effectively until resistance (as a result of excessive particulate loading) diminishes the exhaust capacity of the unit.

Dismantling the System

As gross removal nears completion, filters should be checked for loading and replaced if necessary. If a pre-filter is being used on the outside of the exhaust unit, it should be removed before final cleanup begins. When the negative air system is shut down at the end of the project, the filters should be left in the negative air filtration unit and the openings sealed with polyethylene and duct tape and/or sprayed with spray polyethylene to avoid spreading contamination when the unit is moved from the work site. **Filters in the exhaust system should not be replaced after final clearance sampling is complete in order to avoid any risk of re-contaminating the area.**

Tips for Using Negative Air Pressure Systems:

1. Check the integrity of the gasket between the HEPA filter and housing each time the filter is changed or after the unit has been transported to a new location.

2. A general rule of thumb for filter life during "average" removal is:

   2 hours for the 1/2" pre-filter
   24 hours for the 2" pre-filter
   500 hours for the 12" HEPA filter

   Changing out the 1/2" pre-filter frequently (every 20-30 minutes) during "heavy removal will prolong the life of the much more expensive HEPA filter.

3. Before removal begins, check the availability of a 20 amp circuit. Most negative air machines require 18 amps for startup and 15 amps during normal operation.

4. Negative air units usually pull less volume than the rating assigned by the manufacturer. For instance, a unit rated at 2,000 cfm will typically pull 1300-1500 cfm. Also, as filters load, the cfm is reduced. Note: The reduced flow volume at the maximum accepted pressure drop (see manufacturer's literature) should be the criteria used for this calculation. Adjust your calculations accordingly for the number of units necessary.

5. Start the negative air system before beginning work and check to see if it is functioning properly. Make sure there is adequate makeup air; otherwise the polyethylene may be pulled away from the walls.

6. Smoke tubes are useful for checking airflow inside the containment.

7. Use heavy duty extension cords to energize the negative air filtration units. If a series of cords are connected, take necessary precautions to avoid shock hazards. Make sure the temporary electrical system is properly grounded.

8. As a rule of thumb, the containment area should be no larger than 10,000 square feet for efficient use of a negative air filtration system.

9. The negative air system is more effective in reducing fiber concentrations when laborers start removal at the farthest point from the negative air units and work toward them.

10. When venting the negative air filtration exhaust outside a window, a good seal can be formed by placing a piece of plywood with a hole cut for the flex duct in the window and sealing it with duct tape. Another seal can be formed by placing a piece of six mil polyethylenes over the plywood template and cutting a slip in it for insertion of the exhaust duct. Tape is used to seal the space around the slit in the polyethylene and the duct.

11. The use of supplied air respirators will increase the air pressure in the work area. Negative air filtration units should always be used in conjunction with Type C respirators to prevent build-up of positive pressure.

Wet Removal Techniques

EPA regulations that cover the removal of asbestos material (40 CFR, Part 61, Subparts A&B, 1973 with subsequent amendments and revisions) require wetting the material before removal begins and keeping it wet as it is removed, bagged, transported, and disposed of.

Two advantages to the use of wet methods for removing asbestos materials include a reduction in airborne fiber concentrations that are generated during removal and a reduction in the effort required to remove the material.

Wet removal is based on the ability of water to lower the potential for the asbestos- containing material to release airborne asbestos fibers and increase the settling rate of fibers that are released. Airborne fiber concentrations may be reduced significantly by using wet removal techniques rather than dry.

The positive effects of wet removal can be further enhanced by adding a wetting agent to the water. The wetting agent (i.e. surfactant) is a combination of chemicals which aids in the penetration of water into the material and increases the probability of individual fiber wetting. Various wetting agents are available which have been used in the agricultural industry and fire fighting profession for many years. EPA recommends a wetting agent consisting of 50% polyethylene ester and 50% polyethylene ether in a ratio of 1 ounce to 5 gallons of water. This wetting agent is not as effective with materials that contain a high percentage of amosite and crocidolite asbestos because amphiboles (i.e., amosite and crocidolite) do not absorb water.

Removal of Sprayed or Troweled Friable Surfacing Materials from Ceilings

At this point of the abatement project, the work area has been sealed off with at least two layers of six mil polyethylene on the floors and two layers of four mil polyethylene on the walls (see section on Preparation of Work Area). The decontamination unit and negative air filtration units are in place, and the scaffolding, ladders, various sizes of short- and long-handled scrapers and other removal equipment have been brought into the work area.

REMOVAL OF FRIABLE INSULATION MATERIALS

The first step in the removal process is to thoroughly wet the ceiling material with a low pressure mist of amended water. The material should be misted lightly with amended water to initially wet the surface, and then a saturation coat is applied. The material can be wetted using a low pressure pump system or water hose with garden sprayer attached which can mix the wetting agent with the water. A hand pump garden sprayer can be used for small projects. Application with large pump systems or airless sprayers may cause leakage behind the barrier seals resulting in contamination of the walls and floors. Also, the initial impact of water applied with high pressure may cause elevated airborne fiber concentrations, therefore the low pressure and careful technique in application should be used. Time should be allotted between spraying with amended water and removal to provide for maximum penetration into the material. If the timeframe allows, the ceiling material should be thoroughly saturated with amended water the night before removal starts.

(Note: the added weight of the amended water may cause delamination of this material overnight.)

Removal of ceiling material is carried out in two stages -- gross and secondary removal. Gross removal is typically conducted with a three- or four-man team. Two men working from a mobile scaffold with rails remove the friable material using scrapers. Wide blades can be used if the material comes off easily. Workers of approximately the same height should be paired together on the scaffolds. One or two workers on the ground package the moist material in six mil plastic bags or plastic-lined fiber drums before it has time to dry out.

Rubber dust pans, plastic snow shovels, push brooms, and standard house brooms should be used to collect and bag the wet material. Avoid using metal shovels or dust pans that can cause inadvertent tears in the polyethylene floor barriers. The crew that bags the material also repositions the scaffold as needed, relocking the wheels after each move.
If several crews are removing material, it may be more time efficient to designate a "spray" person, who walks from one area to the next, keeping the material on the ceiling and the floor wet and misting the air to maintain low airborne fiber concentrations.

The spray person can also check for damaged floor barriers and promptly repair them.

Bags containing the waste material are processed for waste load-out, either by wet wiping, placing in another "clean" bag, or placing into fiber drums. (See Waste Disposal Requirements Chapter) All bags should be removed from the work area at least by the end of the work day. Removal of bags on a continual basis provides for easier movement (particularly if workers are wearing air- supplied respirators) in the work area.

After removing as much of the sprayed-on material as possible with scrapers, crews begin secondary removal. Depending on the type of substrate (material underneath the friable insulation), various techniques and tools may be required. Common types of ceiling construction to which friable insulation materials may be applied include concrete, 3 coat plaster system, suspended metal lath, concrete joists and beams, metal deck, corrugated steel, steel beam, or bar joist. The surface substrate may be smooth, rough, or pitted and will affect the difficulty of secondary removal. Typically a combination of brushing and wet wiping is used to remove the remaining residue. The rags used for wet wiping should not leave any fabric fibers on the substrate which might be mistaken as visual contamination. High efficiency particulate air (HEPA) vacuum cleaners are also useful for removing "hard-to-get-to" residue.

While crews are working from scaffolds or ladders to remove all remaining residue from the ceilings, workers should also be cleaning material off the polyethylene wall barriers and any stationary objects in the area. Brooms, wet rags, or squeegees are good for this purpose. Secondary removal is finished when all visual contamination is removed from the ceilings. The next phase is final cleanup (discussed in detail in Chapter12, Cleaning up the Work Area).

Removal of Thermal System Insulation from Pipes, Boilers, and Tanks

There is a wide variation in the types of asbestos- containing thermal system insulation used on pipes, boilers, and tanks. Pipes may be insulated with preformed fibrous wrapping, corrugated paper, and a chalky mixture containing magnesia, fiber felt, and insulating cement.

(Note: There are older materials labeled "magnesia" that contain asbestos and new materials also labeled "magnesia" that contain glass fiber rather than asbestos.)

Usually a protective jacket, which may also contain asbestos, made of paper, tape, cloth, metal, or cement covers the insulation materials. Boilers and tanks may be insulated with asbestos "blankets" on wire lath, preformed block, or the chalky magnesia mixture which is typically covered with finishing cement. Different approaches are typically required for removing these asbestos-containing materials than sprayed-on or troweled-on ceiling insulation. However, the same protective measures are used. Careful handling and packaging is required in many cases because of the metal jackets, bands, or wire associated with the insulation materials.

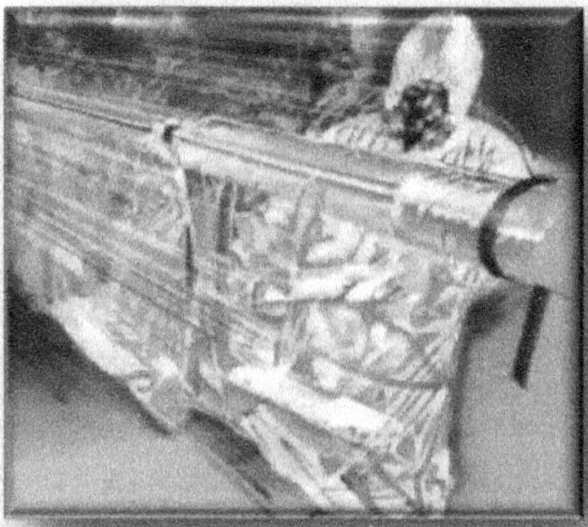

Glovebags, which can be sealed around sections of pipe to form "mini-containment areas", may be used in some situations for removing pipe insulation (see Glovebag Chapter). Insulated objects which are not readily accessible or are too large or hot for application of the glovebag technique, may require a full area enclosure with modified removal techniques.

Because insulation on pipes, boilers, and tanks may contain as much as 70% asbestos and, because areas where these materials are being removed are often confined, high airborne fiber concentrations may occur. Also, these materials are more difficult to saturate with water and may often contain amosite, which is not controlled as well with water as other types of asbestos. If these situations cannot be controlled by higher air flow rates and other engineering techniques, then Type C airline respirators are recommended for workers engaged in removal of asbestos from pipes and boilers.

Removal of insulation from pipes, tanks, or boilers can be accomplished by two-person teams. Cuts or slits are made in the insulation material, a spray nozzle is inserted, and the material is wetted to the extent feasible. One man cuts away the insulation and bags it while the other continuously sprays the material with amended water. Any metal bands or wire that is removed should be folded or rolled and placed in polyethylene to avoid lacerating personnel.

After the gross material is removed, nylon brushes are used to thoroughly clean the pipes, tanks, or boilers. (In cases when pipes are extremely hot, nylon brushes may melt and wire brushes may be the only tool available.) Particular care must be taken to clean the fittings and joints where a cement-plaster type material has been removed. After brushing, the surfaces are wet-wiped and the final cleanup phase begins.

Dry Removal Techniques

NOTE: (Many states will require approval in addition to EPA approval when using dry removal techniques)

Dry removal, which requires specific EPA approval, may be appropriate for some types of asbestos-containing materials that have been previously encapsulated and will not absorb amended water. There are special conditions that preclude the use of water, such as a room containing electrical supply lines that cannot be de-energized during the removal project, hot steam pipes, crawl spaces, etc. Dry removal techniques can be used successfully but require much skill and attention to critical details in order to minimize airborne fibers in the workplace and to adequately confine all airborne fibers to the workplace enclosure. Proven procedures include use of large vacuum systems; small area containment with localized HEPA filtered exhaust, and re-circulating HEPA units inside the work area.

The dry removal procedures selected for a given situation must be carefully matched to the existing work area conditions, the type of asbestos and the skill of the work force. Additional measures could include: adding layers of enclosure plastic; adding airlock chambers to the decontamination units; providing double or triple, rigid primary barriers (in addition to several layers of primary polyethylene); and increasing the number of negative procedures. These added confining and minimizing measures obviously add cost to the project.

It is always much easier to control airborne fibers using wet techniques. It is recommended that all reasonable and safe avenues for wet removal be thoroughly explored before resorting to dry removal. It must also be noted that dry removal requires **job specific EPA approval**, and approval is sometimes difficult to obtain.

It is very important that all personnel use maximum personal protection during dry removal because of the constant and high potential for elevated airborne fiber levels.

Special Considerations

Amended water is not totally effective in controlling fibers emitted from material containing amosite asbestos. Some contractors reportedly use ethylene glycol, removal encapsulants, and/or oils to help reduce amosite emissions.

Others have encapsulants

which are diluted so that it dries slowly and does not harden before the asbestos material can be removed from the pipes or boilers. Some manufacturers are currently conducting comparative testing of these wetting methods to determine which is the most effective.

Steam or hot water distribution networks should be shut down if at all possible, when insulation is being removed.

If these systems must stay on line, special consideration must be given to avoid heat stress and skin burns.

When airline respirators are being used by workers, care must be taken not to let the airlines come into contact with hot pipes which must burn a hole in the rubber line. When air lines are worn by persons working from scaffolds, care must be taken not to wrap the airlines around objects on the ground or the scaffold. See Chapter 5 on Type C Respirators that address safety considerations.

Objective: To provide an overview of non-asbestos related safety and health problems encountered during asbestos abatement projects and provide information necessary to manage these problems.

Learning Tasks: Information in this section should enable participants to:

- Identify, eliminate, avoid, or safely work around potential electrical safety hazards.
- Become familiar with proper procedures and equipment used during asbestos abatement to avoid hazardous conditions and work practices.
- Identify, eliminate, avoid, or safely work around potential fire/life safety hazards.
- Establish effective emergency action plans/procedures specific to the abatement project.
- Identify and eliminate hazards associated with ladders, scaffolds, walking, and working surfaces.
- Become familiar with the basic provisions of OSHA's Hazard Communication Standard.

INTRODUCTION

Asbestos abatement projects have become increasingly technically sophisticated as the body of knowledge grows regarding effective control methods. A great deal of attention has been given to protecting workers and confining fibers. The extra burden of dealing with the asbestos hazard can easily create situations where the basic and more immediate safety hazards can be overlooked.

Safety hazards can manifest if good work practices are not followed. Potential hazards include: electrical considerations, ladders and scaffolding, working surfaces, fire considerations, heat related disorders, and body protection. The methods used in a typical abatement project (sealing the work area, using wet methods, working at heights on ladders and scaffolding, and shutting down normal building systems) add new dimensions to the task of providing a safe working environment.

ELECTRICAL SAFETY CONSIDERATIONS THE HAZARD

One of the most common safety hazards, and one that gives the least warning, is electrical current. Incorrect wiring, improper grounding, and lack of proper shielding results in approximately 1,000 people per year being electrocuted nationwide. Many of these fatalities result from contact with only 120 volts a.c. Three factors determine the severity of electrical shock.

These are:

- The amount of current flowing through the body;
- The path of the current flowing through the body;
- The time the current is allowed to follow this path.

These factors vary greatly. The path of the current depends upon the points of contact. Most often the path is from the hands, through the body, and out the feet. The amount of electrical resistance determines in part the amount of current flow. Moist skin or damp conditions greatly reduce electrical resistance and significantly increase a person's risk of serious injury if he comes in contact with a current source. In addition to the obvious shock potential, many deaths result from falls after a non-fatal electrical shock.

Pre-Work Considerations/Identifying the Hazards

During the pre-bid inspection, preparation of the work site, and during asbestos removal, there are potential electrical hazards that can be identified and eliminated. Examples include:

- Identification of wiring faults in the building: These include open ground paths, reverse wiring polarity, and hot-neutral or hot- ground wires reversed. These common faults can easily be identified with a volt-ohm meter or with plug-in type circuit testers and should be corrected prior to the start up. This is particularly important if these circuits will be used to provide power inside the removal area.

- Un-insulated or exposed and energized wiring or equipment: Asbestos removal jobs are often part of renovation or remodeling projects.Overhead lighting is often removed for cleaning. Equipment or machinery may have been moved out of the area during the removal job and wiring left in place. Damaged equipment or electrical fixtures may not have been repaired by the building owner. All of these things may be combined to create sources of contact with energized electrical circuits. When possible, circuits that will not be used during removal efforts should be turned off and locked out. Wiring and electrical connections should always be considered energized until tested and proven otherwise. Unenclosed wiring junctions in overhead areas are a particularly likely point of contact for removal workers.

- Asbestos abatement projects where the building remains occupied: This is becoming more common as industrial and commercial projects are increasing. This can present problems where electrical circuits or control panels, that are located inside the removal area and that control other parts of the building, must remain energized. Where this situation is encountered, all breakers and switches should be clearly labeled in case power must be secured to other areas of the building during the removal project. Sealing transformers or control boxes may not be possible due to heat build-up. If this situation is encountered, polyethylene will have to be kept away from the surface of the equipment to allow for air circulation. Dry removal may be necessary around energized equipment to avoid a serious shock exposure.

- Providing power inside the removal area: This can create hazards not associated with the building systems. Since OSHA considers abatement projects to fall under the requirements of the 29 CFR 1926 Construction Industry Safety and Health Standards, there are special requirements for supplying temporary power. This may be done by supplying power through Ground Fault Circuit Interrupters (GFCI) or having an Assured Equipment Grounding Program in effect (Many states require the use of GFCIs for electrical safety). Use of GFCIs to protect all circuits provides the safest power source since any significant current leakage will trip the circuit. These devices prove most effective when kept outside the work area away from the high humidity.

- Commonly found electrical devices on abatement projects are: Lights, HEPA vacuum cleaners, negative air systems, drills, saws, heaters, sump pumps, and often, radios. All of these should be inspected regularly for damage, proper grounding, and integrity of insulation.

With the above mentioned items in place, there are still several basic items that should not be overlooked. When possible, scrapers with non-metallic handles should be used for scraping to prevent a possible shock if wiring is cut or contact is made with energized equipment. Insulating the handles of metal scrapers is another option. Hard rubber or plastic scrapers, while more difficult to find, may be used. Fiberglass ladders reduce or eliminate a ground path if a worker contacts an energized circuit.

An assured equipment grounding program requires regular inspection of all tools, cords, and electrical devices along with written documentation.

ELECTRICAL SAFETY REVIEW

- The use of wet methods increases the potential for electrical shock when working around electrical panels, conduit, light fixtures, alarm systems, junction boxes, computers, transformers, etc.

- De-energize as much equipment as possible. Use portable floodlight systems for lighting and regularly check the system and wiring for damage. Twelve volt lighting systems are available that work very well.

- Consider using dry removal in areas immediately adjacent to energized electrical equipment if de-energizing is not feasible. Consultation with local NESHAP authorities is necessary to prevent citation for failure to use wet methods.

- Use non-conductive scrapers and vacuum attachments (wood, plastic, rubber).

- Ensure that all electrical equipment in use is properly grounded before the job starts. This means checking outlets, wiring, extension cords and power pickups. Check for the ground-pin on plugs.These checks should also be made while setting up and regularly during the job.

- Use care not to violate insulated coverings with scrapers, scaffolding wheels, etc. Rolling a heavy cart or scaffold over a flexcord can easily cause internal damage.

- Avoid stringing electrical wiring across floors. Elevate wiring if possible to keep it away from water on the floor and damage from foot traffic and rolling scaffolds. Duct tape is effective for this.

- Do not allow water to accumulate in puddles on work area floors. NESHAP regulations require adequately wet material, not standing water!

- Ensure that electrical outlets are tightly sealed and taped to avoid water spray.

- Always perform a pre-work walkthrough to identify potential sources of electrical hazards to abatement workers, or equipment that may be damaged by wet removal methods. Any miswired or damaged receptacles should be rewired or replaced by a qualified electrician prior to the abatement project start-up.

- Use stable fiberglass ladders – not metal.

- Determine operating voltages of equipment and lines before working on or near energized parts. E-energize and lock-out when possible.

- Electrical equipment and lines should be considered energized unless tested and determined otherwise.

- Energized parts must be insulated or guarded from employee contact and any other conductive object.

- Extension cords used with portable electric tools and appliances must be the three-wire grounded type, designed for hard or extra hard use (Types S, ST, SO, STO, SJ, SJO, SJT, SJTO), and connected to a GFI (Ground Fault Interrupter) circuit.

- Extension cords:
  - ✓ Should be protected from accidental damage.
  - ✓ Should not be fastened with staples, hung from nails, or suspended by wire (tape is an acceptable alternative).

- Portable electric hand-tools should meet the following requirements:
  - ✓ Should be equipped with a 3-wire cord having a ground wire permanently fixed to the tool frame; or
  - ✓ Should be of double-insulated type and labeled as such.

- For circuits over 600 volts, if electrical disconnects are not visible and open or locked out, the following requirements should be met:

  - ✓ Circuits to be de-energized are clearly identified and isolated from all energy sources.
  - ✓ Notification received from a designated employee that all switches and disconnectors that could supply energy have been de-energized, locked out, and plainly tagged to show men at work.
  - ✓ Visual inspections and tests made to assure de-energizing of lines and equipment.
  - ✓ Protective grounds applied to disconnected lines or equipment.
  - ✓ Separate tag and lockout attached for each crew requiring de-energizing of same line or equipment.
  - ✓ Tags and lockouts should not be removed from completed work until designated employees report that all crew members are clear and protective grounds they installed have been removed.

LADDERS/SCAFFOLDING/WALKING – WORKING SURFACES (INSPECTIONS AND PROPER USE)

LADDERS AND SCAFFOLDS

Asbestos abatement projects always present risks to workers from falls, slips, or trips. The nature of the tasks necessitates the use of scaffolding and ladders.

Ladders

The following items should be checked on a regular basis:

- Ladders are always contained in good condition.
- Complete inspections are done periodically.
- No improvised repairs are made.
- Defective ladders are not used.
- Safety feet spreaders and other components of ladders are in good condition. (Missing safety feet create sharp edges that will cut polyethylene floor covers.)
- Movable parts operate freely without binding or undue play.
- Rungs are kept free of grease or oil.
- Ladders are not used for other than their intended purpose. (Ladders should not be used as a platform or walkboard.)
- Extension type ladders should be used with a 1-4 lean ratio (1 foot out for every 4 feet of elevation).
- Step ladders should only be used when fully opened.
- The user faces the ladder while going up and down.
- Tops are not used as steps. If needed, get a longer ladder.

- Bracing on the back legs is not used for climbing.
- Portable ladders are used by one person at a time.
- Ladders are secured to prevent displacement during use.
- All ladders have well-designed safety shoes.
- Hook or other type ladders used in structures are positively secured.
- Fiberglass ladders should be selected to avoid electrical hazards of metal ladders.

Scaffolding

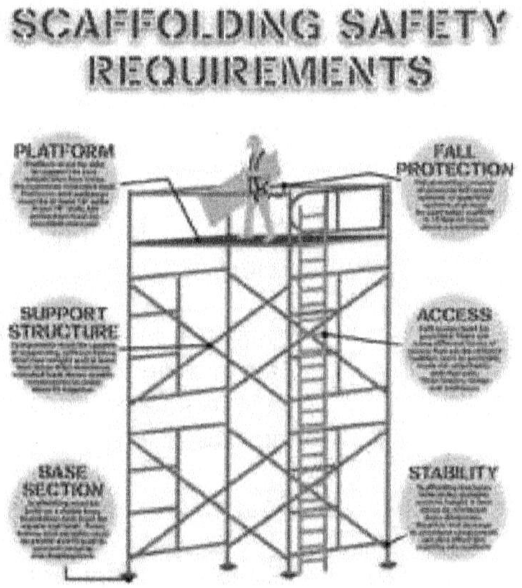

Most asbestos abatement projects will involve the use of scaffolding. Proper setup, regular inspection, and basic maintenance should not be overlooked. In many removal projects, manually propelled mobile scaffolding provides a convenient and efficient work platform. OSHA standards require that when free-standing mobile scaffolding is used, the height shall not exceed four times the minimum base dimension.

This requirement is based on the fact that scaffolding is easily turned over. A simple method to estimate a reasonable amount of force necessary to tip a mobile scaffold if workers try to move it while standing on it. Since relatively little force is required to tip a scaffold, it becomes important to make sure that wheels on mobile scaffolds move freely and are in good repair.

If rented scaffolding is used, all components should be inspected prior to accepting it. Wheels should turn freely and be lubricated. All components such as cross bracing, railings, pin connectors, planking or scaffold grade lumber should be available before the units are assembled. When workers will be riding mobile scaffolding the base dimension should be at least one half of the height.

Workers should be careful to keep debris bagged and obstacles off the floor where mobile scaffolds will be used.

If a wheel catches debris on the floor when the unit is moved, additional force will be required to move it. This additional force may be all that is needed to turn the unit over.

Guardrails should always be installed on scaffolding used for abatement projects. Workers are usually looking up while working and can easily step off the edge of an unprotected scaffold. OSHA requires that guardrails be used when scaffolding is from 4 to 10 feet tall and less than 45 inches wide. Scaffolding 10 feet or higher must have guardrails.

Planking used on a scaffold should not extend farther than 12" over the edges and should always be secured to the frame with cleats on both ends.

Slips, Trips, and Falls

Areas sealed with polyethylene and kept damp to reduce airborne fibers become very slick. Disposable booties are a potential trip hazard; air and electrical lines also create trip hazards. All of these conditions create potential worker hazards even before removal begins. When asbestos and other debris are removed, the accumulations should be bagged and removed from the floor as slick. Disposal booties are a potential trip hazard. All these conditions create potential worker hazards even before removal begins. When asbestos and other debris are moved, the accumulation should be bagged and removed from the floor as soon as possible. This simple step, which may require more initial effort, will make cleanup easier and the overall job far safer. The National Safety Council estimates that there are 200,000 to 300,000 disabling injuries in work-related falls each year. Over 40 percent of the workers are employed in the construction industry.

To summarize:

- Consider the height of the work, equipment in use, and numerous trip hazards. Take a look at your "walking surfaces.
- The use of disposable booties may be impractical in many removal situations. They may come apart and create a serious trip hazard. Seamless rubber boots, slip-on shoes, or safety shoes with non-skid soles may be an alternative depending on the job.
- Inspect ladders and scaffolding for condition.
- Ensure that railings are adequate on scaffolds.

- Minimize water on floors. Wet polyethylene is very slick and water increases the risk of electrical shock.

- Use care around air lines and electrical cords.

- Suspend electrical lines and cords, when possible, using tape.

- No running, jumping, or horseplay in work areas should ever be allowed.

- Minimize debris on floors.

- Pick up tools, scrapers, etc.

FIRE CONSIDERATIONS

Fires can create immediate life threatening conditions. Fire prevention/control should be given a high priority both during planning and removal stages of asbestos projects. A few of the fire safety features to be concerned with are exits, travel distances, emergency lighting, and alarm systems.

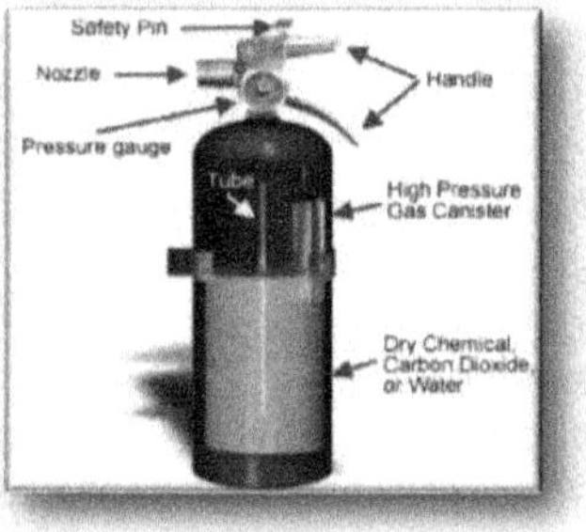

Sealing off an area and blocking entrance/exit openings conflict with OSHA, National Fire Protection Association (NFPA), and local fire code requirements. The contract specifications may state "one means of egress through a properly designed decontamination system"; however, emergency plans should be developed to include alternative exits in emergency situations and these must be familiar to all personnel entering the work area.

Perform a pre-work survey to determine potential fire hazards, sources of ignition, hot spots, and location of exits. Coordinate this with the number of workers to be in the area, the square footage, and the types and amount of combustible/flammable materials that will remain on site.

Some protective clothing will burn and melt quickly. It can shrink, adhere to skin and drip as it burns. Heavy black smoke is a combustion by-product. Polyethylene (it's combustible) will start to burn slowly and pick up speed as more heat is generated. It gives off heavy smoke as the fire progresses. Flame spread is slow and steady. Poly also produces toxic gases during thermal decomposition. Workers would not be adequately protected from smoke with respirators used for asbestos work. Sheeting should be kept away from heat sources such as transformers, steam pipes, boilers, etc., that will be heated during removal. (Polyethylene should not be allowed to contact surfaces above 150°F.)

NOTE: "Fire Resistant" and "Fire Retardant" polyethylene are not "Fire Proof"!

To avoid fire problems in asbestos control areas:

- Ensure that all sources of ignition are removed. Be sure that gas and other fuel sources are cut off and that pilot lights in boilers, heaters, hot water tanks, compressors, etc., are extinguished.

- Locate "hot spots." Quite often you will have to drape equipment instead of sealing off to prevent overheating (i.e., computers, terminal boards, switch panels, transformers).

- Cut off supply to steam lines, electric and steam heaters, and radiators. Do not permit the polyethylene to lay against hot surfaces.

- Do not allow lighters, matches, etc., into the work area. Strictly enforce no smoking, eating, or drinking inside the work area.

- When using an oxygen/acetylene torch to cut pipe, etc., post a fire watch with an appropriate fire extinguisher such as pressurized water. The fire watch should remain at the site for at least one half hour after work is completed. Do not use CO_2 extinguishers in confined or enclosed spaces. Dry chemical extinguishers are effective, but the powder is a respiratory irritant.

- When using a cutting torch, know what is on the other side of the wall and below the floor. Use sheet metal or a treated tarp to catch sparks.

- Reduce the amount of flammable/combustible materials inside a space to a minimum prior to hanging plastic. This includes removal of any chemicals, flammable liquids, heat sensitive materials, etc.

- Mark exits from work area and post directional arrows when exits are not visible from remote work areas. This can easily be done using duct tape on the polyethylene walls and barriers. It is recommended that these directional arrows be placed close to the ground to assist workers who may be crawling in smokey conditions to escape a fire.

- Keep trash and debris to a minimum (i.e., tape, poly, bags, lumber, etc).

- If the work area is large and many workers are present, several emergency exits may be needed. Choose exits that are locked from outside but can be opened from the inside. A daily inspection should be conducted to ensure that secondary exits are not blocked.

- Lighting of exits and exit routes should be provided.

- In case of fire, the fire hazard becomes more immediate than the asbestos hazard and workers may need to violate the plastic barriers. This should be covered with workers in the emergency action plan for the job site.

- Be alert for flammable vapors in industrial areas (solvents such as naphtha, toluene, xylol, etc.).This is especially critical in industrial vacuuming operations where vacuum motors are not explosion proof. Compressed air vacuums may be required.

- A telephone should be available on the job site at all times for notification of authorities in an emergency.

- Post local Fire Department and Rescue Squad phone numbers as a backup to localities with 911 service. Advise them of the operations in progress.

- Ensure that you have a monitor outside <u>at all times</u> trained in emergency procedures. Someone should be trained in first aid, and in the treatment of heat stress.

EMERGENCY PROCEDURES

OSHA requires a written emergency action plan and fire prevention plan. These requirements are detailed in 29 CFR 1910.38. Briefly, the essential items of the plans should include:

- The manner in which emergencies are announced.

- Emergency escape procedures and emergency escape routes.
- Procedures for employees who must remain to operate critical plant operations which may take time to shut down.
- Procedures to account for all employees after evacuation.
- Rescue and medical duties.
- Names and/or job titles of people to be contacted for additional information.
- A list of the major workplace fire hazards.
- Names and/or job titles of people responsible for maintenance of fire prevention equipment.
- Names and/or job titles of people responsible for the control of fuel source hazards.

Establish a system for alerting workers of an emergency or other problem that may require evacuation of the work area. A compressed air boat horn provides an effective alarm that can be heard and does not rely on a power source. All persons entering the work area should be familiar with the evacuation alarm signal and primary and secondary exits. A simple floor plan drawing of the work area should be posted to familiarize persons entering the work area with the site and location of exits.

Written emergency procedures should cover procedures to be used in case of the following: fire, which may include heavy smoke conditions; power failure; compressor failure with the use of air-supplied respirators; accident; or employee injury.

SITE EVACUATION PLAN

MEDICAL SERVICES AND FIRST AID

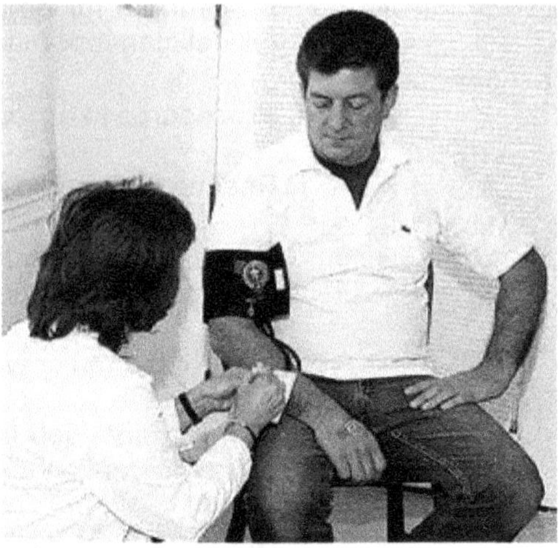

The OSHA Asbestos Standard for the construction industry requires all employees who are exposed to asbestos at or above the action level for 30 or more days per year, or who are required to wear a negative pressure respirator, be given a complete physical examination (OSHA medical screening is required prior to the use of any respirator). The main objective of the examination is to determine that the employee is medically qualified to wear a respirator before performing abatement activities. The examining physician or clinic should be aware that respirators may be worn under hot, adverse conditions. During warm months, heat exhaustion and heat stroke are serious hazards faced by workers, particularly those not acclimated to the heat.

HEAT-RELATED DISORDERS

It is important for the employer to provide training in recognition and awareness of the symptoms and effects of heat stress and heat stroke. It is also important to stress the importance of drinking water and maintaining proper electrolyte levels.

Heat Exhaustion

Causes:

- High air temperature
- High humidity
- Low air movement
- Hard work
- Not enough breaks away from the heat
- Insufficient fluid intake
- Full body clothing
- Workers not acclimated to heat

Symptoms:

- Fatigue, weakness, profuse sweating, normal temperature, pale clammy skin, headache, cramps, vomiting, fainting.

Heat Exhaustion Treatment:

- MEDICAL ALERT
- Remove worker from hot area.
- Have worker lie down and raise feet
- Apply cool wet cloths
- Loosen or remove clothing
- Allow small sips of water or Gatorade™ if victim is not vomiting

Prevention:

- Frequent breaks away from the heat
- Increase fluid intake
- Allow wokers to become acclimatized to heat.
- External cooling (vortex cooling, ice vests)

Heat

Causes:

- High air temperature
- High humidity
- Low air movement
- Hard work
- Not enough breaks away from the heatr
- Not drinking enough water
- Full body clothing
- Workers not acclimated to heat

Symptoms:

- Dizziness, nausea, severe headache, hot dry skin, confusion, collapse, delirium, coma, and death

Treatment:

- MEDICAL EMERGENCY
- Remove worker from hot area
- Remove clothing
- Have them lie down
- COOL THE BODY (SHOWER, COOL WET CLOTHS)
- **Do Not** give stimulants

Telephone numbers of the physicians, hospitals, and ambulances should be conspicuously posted for emergency use.

Means should be available for prompt transport of an injured person to a physician or hospital, and there should be a telephone with emergency numbers available.

Before beginning the project, provisions should be made for prompt medical attention in case of serious injury or other medical emergency.

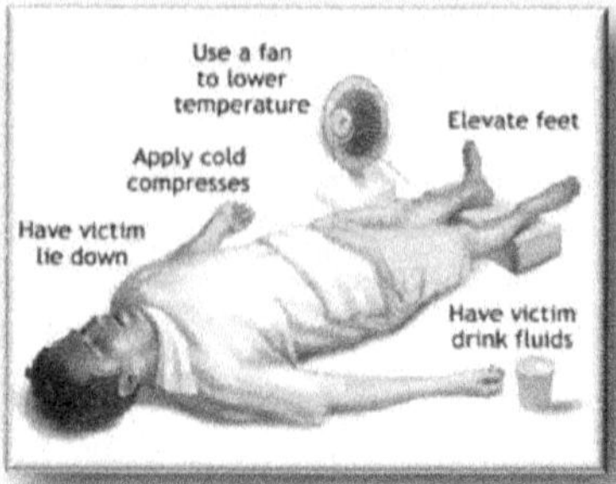

Someone trained in basic first-aid should always be on the abatement project.

BODY PROTECTION

The following guidance should be used when addressing whole body protection for abatement area personnel:

Provide and require use of special whole body clothing, including shoes, for any employee exposed to airborne concentrations of asbestos.

- Provide work gloves as part of whole body protection to employees exposed to asbestos. This is particularly important when metal lath, suspended ceiling grids, and other materials are being removed.

CARBON MONOXIDE POISONING

When airline respiratory protection is used, it is important that the outside monitor be familiar with the system and any problems associated with breathing air. Carbon monoxide poisoning is perhaps the most important of these problems. It is important to note that these symptoms are similar and may be confused with those from heat stress.

Description of CO: Colorless Odorless, and Tasteless

Sources

- Oil-lubricated compressor
- Internal combustion engine
- Open flame and fire
- Unvented gas
- Kerosene heaters

Symptoms:

- Dizziness, nausea, headache, drowsiness, vomiting, collapse, coma, and death (note similarity of symptoms to heat stroke)

Limits:

35 ppm (Time weighted average over 8 hours) OSHA
200 ppm ceiling limit – OSHA

10 ppm Grade D breathing air for airline respirators) (Maximum allowable concentration) (Note: ANSI/CGA G7.0 – recommends 10 ppm maximum.)

If these symptoms are observed, those persons should immediately brought into fresh air and medical attention should be provided.

Monitor any prescription or over the counter medicines being used by employees. These may cause an adverse reaction when used by persons under strenuous conditions common to removal work.

- Scrapers, retractable knives, wire cutters, chisels, and other sorts of bladed tools are frequently used. Always cut away from the body. Provide tools with insulated handles.

- Many puncture and cut wounds occur when removing metal lath or cutting duct work. Use care and have a good first aid kit available.

- Protective hardhats must be worn on a jobsite where there is exposure to falling objects, electric shock, or burn.

- Provide, require the use of, and maintain in sanitary and reliable condition, protective equipment necessary to protect any employee from any hazard that could cause injury.

- Wear face shields or goggles for operations involving potential eye injury. Full face respirators are most effective.

- Check with your surfactant supplier on irritant properties of your wetting agent. (Always have a material safety data sheet on all of your materials and familiarize workers with any cautions or special considerations for their safe use.)

- Arrange work so workers do not have to reach extensively overhead. Get them up to the job!

- Instruct your workers on proper lifting methods. Nothing will take the profit out of a job faster than a serious back injury.

- Use the “buddy system” for lifting and moving heavy objects.

- Use hand carts or rolling pallets when possible. Keep manual material handling to a minimum.

MISCELLANEOUS

OSHA requires that a poster be permanently posted on the job site notifying workers of their rights under the act. This poster, commonly known as the "Job Safety and Health Poster," is available from OSHA offices.

When an employer has more than 10 employees at any time during the calendar year, he is required to maintain a record of injuries and illnesses that occur. Part of this requirement may be met by filling out accident reports required by Worker's Compensation insurance carriers. The other requirement is maintenance of the "Log and Summary of Occupational Illnesses and Injuries – OSHA Log 300 and 300A." These forms and a booklet titled "What Every Supervisor Needs to Know About OSHA Record-keeping," are available from OSHA and provide information on these record keeping requirements.

HAZARD COMMUNICATION STANDARD 29 CFR 1926.59

OSHA has expanded this standard to include the construction industry. The purpose of this standard is to ensure that the hazards of chemicals or materials used in the workplace are identified and that this information, along with information on protective measures, is passed on to employers and employees. Elements required under this standard include:

- Comprehensive written hazard communication program;
- Labeling of hazardous materials;
- Maintaining material safety data sheets;
- Employee training.

Employers are required to inform affected workers about hazardous chemicals they may be exposed to through:

1. A written Hazard Communication Program which must include:

    a. Plans to meet the criteria of the standard relating to the labeling, material safety data sheets, and employee training.

    b. A list of all hazardous chemicals/materials.

    c. The methods to be used to inform employees and outside contractors of hazards of non-routine tasks.

    d. The hazards associated with chemicals contained in unlabeled pipes or vessels in the work area. This also applies to hazardous materials released while using a product.

    e. The methods to be used to inform outside contractors who may work on the premises of the hazards to their employees.

2. Material Safety Data Sheets. All chemicals used in the workplace must have material safety data sheets available which must include all health hazard exposures, as well as physical hazards and emergency procedures. All material safety data sheets must be accessible by all employees during any working time, which includes all three shifts as applicable.

3. Labels. All containers in the workplace must be labeled, marked, or tagged with the identity of the hazardous material contained and the appropriate hazard warning, and the name and address of a responsible party.

4. Employee Training. Employees exposed to any hazardous chemicals in their work environment must be educated as to:

   a. The requirements of this standard.

   b. The operations involving hazardous chemicals, the location and accessibility of the material safety data sheets, and the location and content of the written hazard communication program.

   c. The methods and observations that may be used to detect hazardous chemicals.

   d. The physical and health hazards of the chemicals in the work area.

   e. Measures employees can take to protect themselves.

   f. How employees can obtain, interpret, and use the information in the written hazard communication program.

It is recommended that you refer to the actual standard for more detailed information.

Exposure to hazardous materials can occur in a number of tasks associated with asbestos abatement work. Examples may include: spray adhesives; surfactants; encapsulants; paints or other products used for lockdown of fibers; materials to be left in the work area; and of course, the asbestos. Training required in the asbestos standard will have to be expanded to include other hazardous materials on the job.

Information on possible hazardous exposures should be reviewed with employees before the exposure occurs so that proper precautions can be taken. Material safety data sheets are available from manufacturers, suppliers of products, and from owners of buildings where hazardous materials are handled in the removal area. Contractors fall under the umbrella of hazard communication programs of building owners who work with hazardous substances when contractor personnel may be exposed to those materials.

Objectives: To provide instruction on effective techniques for cleaning up the asbestos abatement work area, addressing initial gross cleanup through final wipe down.

Learning Tasks: Information in this chapter should enable participants to:

- Become familiar with materials and equipment used to accomplish cleanup
- Understand the basic procedures for conducting specific cleanup tasks.
- Recognize the importance of proper sequencing of cleanup tasks.
- Understand what to look for during a visual inspection.
- Perform proper cleanup during and after gross removal is complete.

CLEANING UP THE WORK AREA

Although cleanup is a tedious, sometimes lengthy process, it is one of the most critical tasks of the project. Successful cleanup operations require proper sequencing of tasks and great attention to detail. If these items are overlooked, much more time may be spent in the re-cleaning and air monitoring cycle than would have been spent to initially conduct a thorough, correct cleanup. Sequential steps and details for cleaning up an area where sprayed-on material has been removed are provided in the following discussion. Removal and cleanup operations for thermal system insulation in areas such as boiler rooms may vary, depending on individual specifications and the nature of the project.

CLEANING DURING GROSS REMOVAL

Cleaning of the work area begins concurrently with the removal of asbestos-containing material from the substrate. Techniques for limiting fiber release and cleaning the work area during the removal phase are addressed in the chapter "Confining and Minimizing Airborne Fibers." In summary, a floor crew wearing appropriate personal protective equipment is responsible for bagging the material soon after it is removed, while it is still damp.

The material is collected from the floor with brooms, squeegees, plastic dust pans, or other appropriate tools and placed in six mil labeled bags for disposal. Metal shovels and other sharp objects should be avoided since they will cut and tear the floor polyethylene sheets.

FINAL CLEANUP

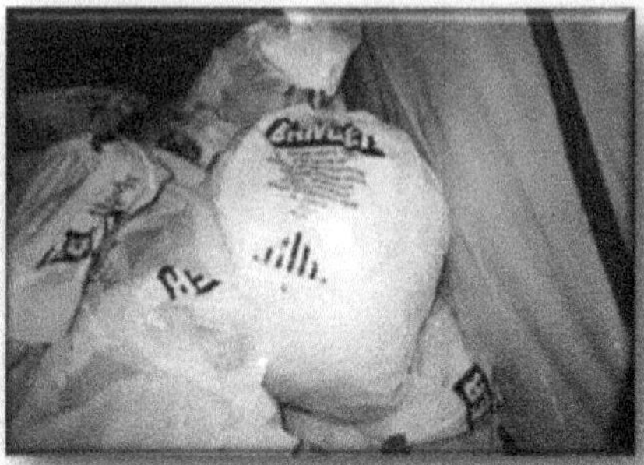

The discussion on final cleanup applies to the phase of the project in which all visible asbestos-containing material has been removed from the substrate and the substrate has been brushed (with nylon bristle brushes) and wet wiped. A flow chart recommending a good and feasible sequence of tasks for performing cleanup is provided at the end of this chapter.

Remove Gross Contamination From Wall Covering or remove inner contaminated Layer

After all visible asbestos-containing material has been removed from the substrate, the next cleaning task should be the removal of all visible asbestos contamination which has splattered or collected on the polyethylene wall coverings. Preferably, two layers of polyethylene were initially hung on the walls, and the inner contaminated sheets can be removed at this point instead of cleaned. Ideally, the contaminated sheet is lightly misted with an encapsulant or "lockdown" material (see "Post Removal Lockdown Procedures and Asbestos Substitute" chapter) to minimize the release of airborne fibers. After detaching or cutting this first inside layer of polyethylene from the bottom of the wall, workers should use ladders to reach the top of the wall sheet. The inner sheet of the work area enclosure should be gently detached from the top of the wall and folded inward to form a compact bundle which can be packaged in a properly labeled six mil polyethylene bag for disposal. Any visible debris that leaked behind the inner layer of polyethylene onto the outer (final) layer is now removed with a HEPA vacuum and/or wet wiping methods.

Remove Gross Contamination from Equipment in Work Area

The next cleaning efforts should be directed toward removing gross contamination from the exteriors of the negative air filtration units, scaffolding, ladders, extension cords, hoses, and other equipment inside the work area. Cleaning can be accomplished using a combination of HEPA vacuuming and wet wiping. This is also a good time to change-out any of the filters that need replacement on the negative air filtration units.

Remove Top Layer of Floor Polyethylene

At this point, the top layer of six mil poly, which has been used to cover the floor area, should be cleaned appropriately and carefully folded inward to form compact bundles for bagging and disposal. Any visible contamination that leaked through to the bottom floor layer should be removed (i.e., HEPA vacuumed, wet wiped). This material should then be bagged and disposed of according to procedures outlined earlier. Excessive time should not be spent in cleaning the floor sheets, but any obvious contamination should be cleaned, contained, and disposed of properly.

Conduct Visual Inspection of All Surface Areas/Re-clean if Necessary

After all of these cleaning tasks have taken place, a thorough visual inspection of the area should be conducted. The inspector (building owner's representative) and the contractor's representative (usually the project supervisor) should check for visual contamination or residual debris on the substrate from which the asbestos-containing material has been removed. Ledges, indented corners, and other surfaces that might "catch" falling material or contain residual material must be inspected closely. A high-intensity flashlight will prove helpful during this inspection. As the building owner's representative and the contractor's representative walk through the work area, the inspection and re-cleaning process might be facilitated by recording on paper the items or areas that need additional cleaning. The contractor's representative is responsible for correcting any of the deficiencies noted during the inspection before beginning the next phase of work. The American Society for Testing and Materials (ASTM) has issued ASTM E1368 "Standard Practice for Visual Inspection of Asbestos Abatement Projects," which presents guidelines for conducting visual inspections following an asbestos response action.

Perform Final Wipe Down of Equipment/Remove from Work Area

After the work crew has completed re-cleaning the areas noted on the inspection list, equipment in the work area should be thoroughly cleaned (gross contamination was removed earlier). Equipment should be wet wiped, washed off in the shower at the waste load-out area, wrapped in polyethylene, or placed in plastic bags. Tools such as scrapers, utility knives, and brushes can be placed in buckets or pans (bottoms cut off of fiber board drums work well) and then sealed in plastic bags for transport to the next project. Equipment that is not needed for completion of the project should be removed from the work area. The negative air filtration units remain in place and operating for the remainder of the cleanup operation until final clearance samples are collected.

Remove Polyethylene Floor Covering/Remove or Clean Carpet

After vacuuming of these areas is completed, the polyethylene floor covering is detached from the wall and folded inward to form a compact bundle for bagging and disposal. If a carpet is still in the work area and specified for removal (removal instead of cleaning is the preferred practice), workers should lightly mist the entire carpet with amended water before detaching it from the floor and rolling it up. Once the carpet is rolled up, it can be wrapped with six mil poly, sealed with duct tape, and properly labeled for disposal. A note of caution: In some locations, carpet may be stuck to the floor with glue that does not readily separate from the flooring. As the carpet is taken up, some portions of the backing may tear away and remain stuck to the floor. Several unplanned additional man-hours may be required to pry or scrape up the glue-carpet spots that are left after the carpet is removed. Also, tearing of the carpet material may elevate fiber counts in air samples analyzed by phase contrast microscopy. If the carpet is not specified for removal, at least one layer of polyethylene should be left in place to protect it from residual contamination. Note: In many cases, it will be easier to remove carpet following preparation of the work area and prior to gross removal of asbestos-containing material. If the carpet remains in the work area and becomes saturated, it may be very difficult to deal with.

HEPA Vacuum

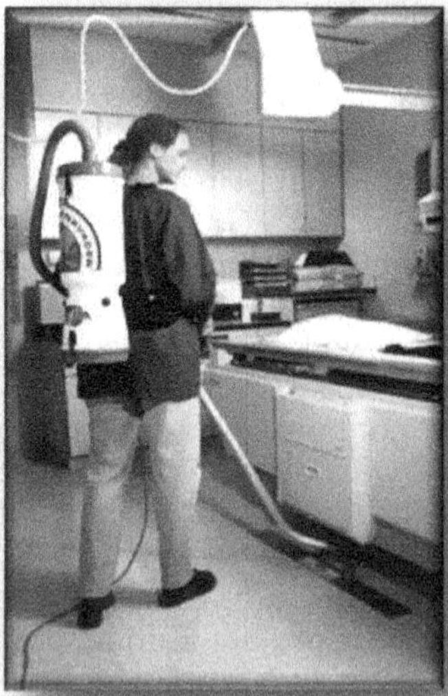

After the floor area is uncovered, corners and crevices can be cleaned with a HEPA vacuum. It may also be necessary to wet wipe certain locations.

Wet Wipe Walls

At this point, the walls and floor polyethylene (if applicable) are HEPA vacuumed and/or wet wiped. (If carpet has been removed, the floor should be thoroughly cleaned.) Workers should begin cleaning the areas farthest away from the negative air filtration units and use amended water to wet wipe all exposed surfaces (excluding the substrate from which the asbestos material was removed). For best results, workers should use cotton rags or lint free paper towels, which are disposed of after one use. Rinsing and reuse of towels may result in smearing asbestos fibers on surfaces and not contribute to overall cleaning. Also, to avoid smearing of residual fibers, workers should wipe surfaces in one direction only.

Paper towels should not be used to wipe down rough surfaces and should be discarded before they begin to deteriorate when used on smooth surfaces. Small "fibrous looking" residue, which may be deposited on surfaces as a result of using deteriorated paper towels, could cause a problem during the final visual inspection.

Wet Mop Floors

After the walls are wet wiped, the floor (or the floor polyethylene) is mopped with a clean mop head wetted with amended water. (Caution: If carpet remains in place, only minimal amounts of water should be used during this process.) The water should be changed frequently.Waste water from the wet wiping and mopping operations should be treated as asbestos-containing waste and dumped in the shower drain to be appropriately filtered or placed in a barrel for disposal.

Wait Overnight/Repeat Wet Wipe and Wet Mop Procedures

Frequently, abatement project specifications will call for "3-phase cleaning." This may require more time spent on the project, but if properly conducted will actually save money and prevent confusion at the conclusion of the project. After the walls and other surfaces (shelves, ledges, etc.) have been wet wiped and the floors have been mopped, activity in the area may be stopped until the following day. The next day, the same wet wiping and mopping procedures are often repeated. As an alternative to using amended water for the second wipe down, the cleaning efficiency may be increased by using a commercial cleaning product such as Endust™ or Pledge™.

Visual Inspection/Re-clean if Necessary/Re-inspect

Work areas should be dry before the final visual inspection is conducted. The inspection is again conducted by the owner's representative and the job supervisor. All surfaces are carefully checked for visible contamination and any areas that need further cleaning is listed on paper. Ledges, tops of beams, and all other hidden locations should also be inspected for asbestos-containing dust or debris at this time. After any necessary re-cleaning has been conducted, the inspector and job supervisor make a final walkthrough to assure that the items listed have been addressed. Here again, the ASTM standard for visual inspections may be used as a guideline for final inspections.

Following this second visual inspection, the lockdown of any microscopic asbestos fibers should be completed. This procedure is covered in detail in the chapter, "Post-Removal Lockdown Procedures and Asbestos Substitutes."

Clearance monitoring is addressed in detail in the chapter on "Air Sampling Requirements." When the air sampling results indicate that the airborne fiber concentration meets the predetermined criteria for clearance, the final "critical" seals on vents and stationary objects such as water fountains, electrical outlets, etc., can be removed.

If the first set of air samples indicates airborne fiber concentrations in the area are above the specified "clearance level," the area must be re-cleaned followed again by clearance sampling. This cycle is repeated until results of airborne fiber concentrations indicate that the clearance criteria have been attained.

FINAL CLEARANCE MONITORING

Following passage of final air clearance testing, a final visual inspection should be performed by the building owner or his representative and the contractor or his representative. This final visual inspection will look for any gross asbestos debris that may have been trapped behind critical barriers or underneath the decontamination unit. If any debris is discovered, it must be cleaned appropriately.

After the area has been cleared for re-occupancy by unprotected personnel, remaining renovation can be initiated (i.e., painting walls, installing suspended ceiling, or laying carpet).

CLEANING UP THE DECONTAMINATION UNIT

For decontamination units that are not prefabricated or modular, the unit is lined with three layers of polyethylene on the floor and one or two layers on the walls (at a minimum, the walls of the equipment room should be lined with an extra layer of polyethylene). The top layer of floor poly in the equipment room should be removed at the same time the top layer of floor poly in the work area is removed, using the same procedures. This will minimize tracking contamination back into the work area. After cleanup is completed inside the work area, the polyethylene on the walls of the decontamination unit is lightly misted with amended water and folded inward. Next, the remaining layers on the floor are removed in the same manner and packaged with the other poly for disposal. The walls should be visually checked for contamination and wet wiped if necessary. The decontamination unit can now be disassembled for transport.

CLEANING UP THE ENCLOSED TRUCK

During the last disposal trip to the landfill, after the truck has been emptied of all waste materials, the polyethylene lining in the inside of the truck is misted with amended water and carefully removed. Good practice should include wet wiping the floor and walls of the truck at this time. Polyethylene removed from the truck interior and the protective clothing worn by workmen conducting disposal are bagged for disposal and placed with the other materials at the dump site.

SUGGESTED SEQUENCE FOR CLEANING UP AN ASBESTOS ABATEMENT WORK AREA (A SIMPLIFIED SCHEME)

1. Continuous Cleaning During Abatement	7a. Second, More Stringent, Visual Inspection of outer Polyethylene Layer Still in Place
2a. Complete Gross Removal and Initiate Final cleanup	7b. Check for Possible Contamination to Inside Layer
2b. All Asbestos Waste Out of the Work Area	7c. If Asbestos Debris is Discovered During Inspection, Further Cleaning is Necessary
3 Visually Inspect for any Residual Asbestos Work Area	8a. With Outer Layer of Polyethylene sheeting and Critical Barriers Still in place, Pre-Clearance Air Samples may be collected.
4. Lockdown/Encapsulant Solution Applied to Substrate and Inner Layer of Polyethylene Misted to Contain Residual Asbestos Fibers	8b. Pre-Clearance Results Indicate Work Area Clear- Proceed to Step 9
5. Lockdown Material Dry/Setup	8c. If Work Area is still Contaminated-Re clean and Sample Again
6. Inner Layer of Polyethylene Taken Down, Contained, & Disposed of as Asbestos Containing Waste	9. Outer Layer of Polyethylene Removed, Critical Barriers Remain in Place, Final Clearance Air Samples Collected
	10. Work Area Passes Final clearance for Re-occupancy by Unprotected personnel

Objectives:

To provide an overview of correct methods and regulatory requirements for disposal of asbestos-containing waste resulting from asbestos abatement projects.

Learning Tasks:

Information in this chapter should enable participants to:

- Understand correct procedures regarding the disposal of asbestos-containing waste.
- Become familiar with procedures of notifying the appropriate agencies.
- Understand the appropriate labeling requirements, wet methods, and packaging procedures.
- Know requirements for effective transportation of asbestos-containing waste and disposal at the landfill or disposal site.
- Become familiar with appropriate OSHA and EPA regulations regarding asbestos waste disposal.
- Understand requirements for record keeping.

PREPARATION OF ASBESTOS-CONTAINING WASTE BEFORE TRANSPORTATION TO THE DISPOSAL SITE

Wetting

Once the asbestos-containing waste material has been removed from the areas of concern, certain precautions must be taken before disposal operations begin. The first, and probably most important, undertaking is to ensure that all of the asbestos-containing waste has been thoroughly treated with water, or "wetted." This may be accomplished by having a water supply available in any area that abatement work is taking place (i.e., low-pressure water sprayer). As the asbestos-containing material is being removed, the material should be kept adequately wet via a low pressure water stream. By ensuring this, the chances of airborne asbestos fiber generation are significantly reduced. Wet waste material will then be suitable for containerizing. Wet waste material will then be suitable for containerizing.

Containerizing

An effective way to ensure that the asbestos-containing waste has been properly packaged for transportation to the disposal site is to establish a standard procedure for bagging and handling the waste. The first step in this procedure would be to obtain required 6-mil polyethylene bags (Most states require that all asbestos waste to be double bagged or wrapped in 2 layers of 6 mil poly and accurately labeled). These will be air-tight and puncture resistant, and must be properly labeled.

After the proper bags are selected, the next step is to train the abatement workers in the proper techniques for containerizing the waste materials. Important concepts of this training should include:

a. Discussion of the importance of handling asbestos-containing waste in a careful manner to keep airborne fiber generation minimal.

b. Instruction on materials that should not be included in bags (i.e., metal, sharp objects) and also that each bag should be considered "full" when it is half filled (since material saturated with water will be heavier).

c. Instruction on correct procedures for sealing off waste-containing bags with duct tape. Ensure that all excess air is squeezed out of bags before they are sealed (to conserve space).

d. Discussion on the importance of ensuring that the asbestos warning label on each bag is legible, so that no bags will be disposed of mistakenly.

Once the asbestos-containing waste is securely enclosed inside the bags, the best recommended practice is to decontaminate the bags by wet wiping or HEPA vacuuming them clean. The bags may then be placed in fiberboard or metal drums with locking rims. These drums should be labeled in the same manner as the bags. To use the drums efficiently several bags can be placed in each drum. Important concepts that should be included when instructing workers in drum usage are:

a. Prior to the time drums are to be used, while they are still in the waste load-out area, an effective method of contamination control is to line the outside of each drum with a plastic (garbage) bag. This outside bag should be kept on the drum while it is being filled with the asbestos-containing waste bags.

b. Once the drum is filled, the lid or rim should be locked into place. The drum will then be ready for transportation out of the work area.

c. Before leaving the work area (at the doorway to the waste load-out area), the plastic bag on the outside of the drum should be removed and placed in the next drum to be filled with waste.

d. Before the drum enters the load-out area, it should be hosed down and/or wet wiped to ensure that there is no residual contamination present on the outside of the drum.

e. Immediately after this bag transfer is accomplished, the sealed drum should be moved into the waste load-out area and subsequently into the enclosed truck.

Drums may not be used for asbestos removal in some states since many of their landfills will not accept them.)

NOTE: Please see local and state diagrams for waste loadout and decontamination systems. Many states have specific requirements.

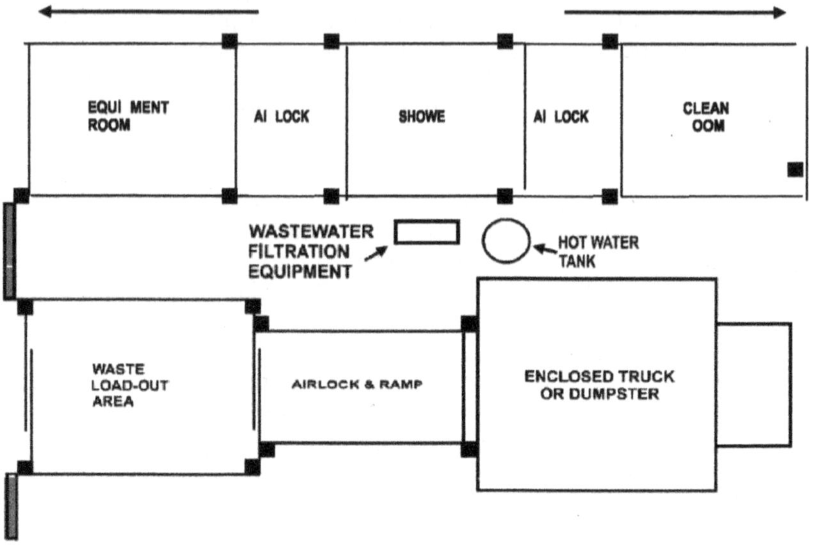

Layout of decontamination unit with equipment room and waste load- out unit openings directly into work area.

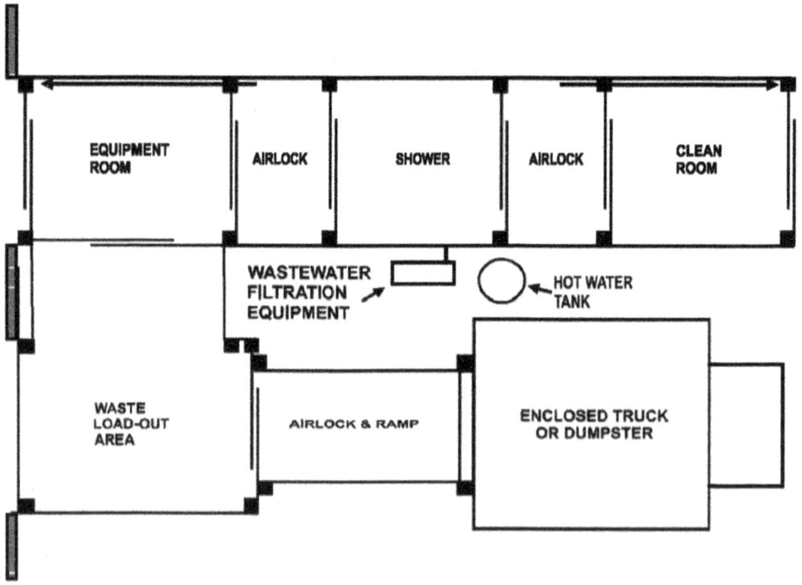

Layout of decontamination unit and waste load-out unit with waste load-out opening into equipment room.

Labeling

It is important that only bags, drums, and wrappings that are properly labeled be used for containerizing, transporting, and disposing of asbestos waste. All drums and bags must be labeled according to the NESHAP regulations, which require the OSHA labeling:

The NESHAP regulations also require that any asbestos-containing waste materials transported offsite from where the waste originated must be labeled with the name of the waste generator and the location where the waste was generated.

The U.S. Department of Transportation (DOT) also has labeling requirements for asbestos waste containers. These labels must state:

US DOT **Class 9** **Packaging group III** **RQ** **WASTE ASBESTOS** **MIXTURE**

The name of the waste generator or waste transporter must also be included. In the above label **RQ is a reportable quantity** (1 pound or more of friable asbestos), WASTE indicates waste material, ASBESTOS is the shipping name, MIXTURE indicates the material is mixed with a binder or filler, and NA2212 is the North American shipping number assigned to asbestos.

WASTE LOAD-OUT PROCEDURE

Inside Team

The most effective method in a waste load-out procedure is to use two teams of workers: an inside team and an outside team. Wearing appropriate respirators and protective clothing, the inside team will be responsible for ensuring that the drums are properly packed, lids locked into place, and plastic bags removed from the outside of each drum before it is sent through the waste load-out area and into the enclosed truck. (The plastic bags should then be placed in the next drum to be disposed of.) It is important that no workers from the inside team exit the work area through the airlock.

In cases where the drums are not being covered with plastic bags, it becomes necessary for the inside team to assure that each drum exiting the work area be free of any dust. This may be accomplished by inspecting and wet-wiping every drum leaving the area.

Outside Team

Wearing dual-cartridge respirators and appropriate protective clothing, the outside team (in the waste load-out area) will post themselves at the entrance to the work area.

They will receive the drums into the load-out area from the inside team. Then, the outside team will load them into the enclosed truck. The entrance into the waste load-out area from the work area should be secured to prevent any unauthorized entry or exit.

Drums should be placed on level surfaces in the enclosed truck and packed tightly together to prevent shifting and tipping. Under no circumstances should containers ever be thrown into the truck. Also, when moving the containers, hand trucks, dollies, or pull cars should be used. In addition, it is important to instruct workers in proper lifting techniques in order to avoid back injuries. Where ramps are not possible, trucks with lift gates would be helpful for raising drums during loading.

To assure that the truck is properly enclosed, the inside area should be lined with two layers of six mil polyethylene. First, the floor should be completely covered with a six-inch overlap of each piece. The same method should also be used when lining the sides and top of the truck. Duct tape should be used to properly secure the sheets of polyethylene. This will not only ensure additional enclosure of the asbestos-containing waste, but it will also provide for easier clean-up operations.

It should be noted here that any debris or residue observed on containers or surfaces outside of the work area resulting from disposal activities should be immediately cleaned by using HEPA filtered vacuum equipment and/or wet wiping, as appropriate.

OTHER FORMS OF ASBESTOS-CONTAINING WASTE

In any asbestos abatement project, not all of the waste material that needs to be disposed of will be loose or broken apart. There are many cases in which it will be necessary to dispose of materials such as asbestos-containing floor, wall, or ceiling tiles, shingles, rugs, etc. The rule of thumb to follow in these instances is simply good common sense. This may include neatly banding together tiles or shingles, with care not to expose sharp edges or any other protruding objects that could possible puncture the polyethylene enclosure. Once the materials are banded together, each bundle should be wrapped in two layers of six mil polyethylene, secured by duct tape, and labeled appropriately. When this is complete, the bundles should be neatly stacked in the truck. Care should be used so that tipping or shifting of the load is prevented.

TRANSPORTATION TO THE WASTE DISPOSAL SITE

As work progresses, and to prevent exceeding available storage capacity on-site, sealed and labeled containers of asbestos-containing waste should be removed and transported to the pre-arranged disposal location. Regulations may vary from state to state, but there are standard procedures that must be followed in any operation involving asbestos waste disposal. Disposal must occur at an authorized site in accordance with regulatory requirements of NESHAP (National Emission Standards for Hazardous Air Pollutants) and applicable local guidelines. It is best to check with state officials on these requirements.

When transporting asbestos-containing waste to any disposal location, it is important that the drivers of the vehicles be properly trained in correct waste handling procedures. It is also important that they not use excessive speeds or unusually rough roads to avoid load slippage or tipping. The driver will be responsible for retaining all dump receipts, trip tickets, transportation manifests, or other documentation of disposal. These should then be given to the building owner for his/her records.

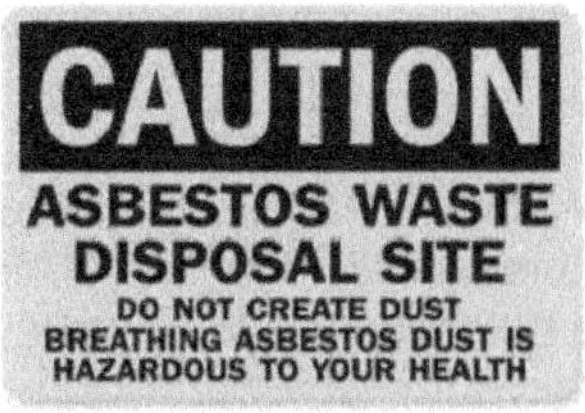

DISPOSAL AT THE LANDFILL

Once the truck containing asbestos waste arrives at the landfill, the driver should approach the disposal location as closely as possible for unloading of the waste materials. Bags should be taken out of the drums along with the other waste components. They should be inspected as they are off-loaded. In the event a bag has been damaged, the material should be repacked in another bag as appropriate. There may be some instances in which the drums can be buried at the landfill; however, it is usually more economical to reuse the drums.

Waste bags should be placed on the ground at the disposal site, not pushed or dropped out of the trucks, as the weight of the wetted material could rupture the containers. Personnel off-loading the containers should wear proper protective equipment, which includes disposable head, body, and foot protection. (Minimum respiratory protection requirements should include the use of half-face, air-purifying, dual-cartridge respirators equipped with high-efficiency filters.)

Upon complete removal of all containerized waste, the truck cargo area should be decontaminated using HEPA vacuums and/or wet wiping methods to comply with the EPA "no visible emission" criteria. The polyethylene sheeting should be removed and discarded along with contaminated cleaning materials and disposable protective clothing in other bags and/or drums at the disposal site. Landfill personnel should have their own personal protective equipment.

Bags or drums should be placed intact in an excavated area and covered with a minimum of six inches of earth at the end of each working day. These areas must be clearly marked to prevent future disturbance of the waste.

The EPA Regional Asbestos Coordinator in the area that the asbestos abatement work is taking place can usually provide a list of approved sites for disposal of asbestos-containing waste. (Ask your instructor for a current list)

RECORDKEEPING

It is very important to keep proper records of waste shipments in order to avoid potential future legal problems. The NESHAP regulations require that the waste generator (either the building owner or contractor) of all asbestos waste disposed of off-site from where the abatement project took place complete a Waste Shipment Record. This record must be provided to the disposal site owner or operator at the same time the waste is delivered to the site.

The Waste Shipment Record must include the following information:

- Name, address, and telephone number of the waste generator;
- Name and address of the local, state, or EPA regional agency responsible for administering the asbestos NESHAP program;
- The quantity of asbestos-containing waste material in cubic yards or cubic meters;
- The name and telephone number of the disposal site operator;
- Name and physical site location of the disposal site;
- Date transported
- Name, address, and telephone number of transporter(s);
- Certification that the waste was properly classified, packed, marked, labeled, and transported.

The landfill owner or operator is required by NESHAP to send a signed copy of the Waste Shipment Record back to the waste generator within 35 days of the date the waste was accepted by the initial waste transporter.

OTHER CONSIDERATIONS FOR ASBESTOS-CONTAINING WASTE DISPOSAL

One aspect that must not be overlooked when devising an asbestos-containing waste disposal strategy is that of filtering the run-off water from showers in the worker decontamination area. It is now generally accepted that filtration of asbestos-contaminated water through a five micron filter is the state of the art for asbestos removal from water. Discharge of the filtered water should be to a sanitary sewer system, or in its absence, to a septic tank and field system with adequate capacity.

If there is any uncertainty regarding water regulations in a particular area, the best course of action to follow would be to contact the state department of environmental management. Local municipalities have the final authority in regard to waste water treatment and should always be contacted for information concerning regulations.

WASTE DISPOSAL ALTERNATIVES

Alternatives to traditional disposal of asbestos waste in landfills are currently being researched and developed in many universities and private companies. The most extensively researched technique for changing asbestos-containing waste material into non-asbestos (asbestos-free) material is glassification of the waste. This is done by heating the material to extremely high temperatures in a furnace or using a plasma-arc torch, which changes the debris into a glass-like substance with no detectable amounts of asbestos. The glass-like material can then be used for construction material.

Although the classification of asbestos-containing waste shows promise, currently, the technique is very expensive, not readily available, and must still address the other demolition or renovation debris usually associated with asbestos containing waste.

EXAMPLE WASTE SHIPMENT RECORD

Generator		
1. Work site name and mailing address	Owner's Name	Owner's Phone No
2. Operator's name and address	Operator's Phone No.	
3. Waste disposal site (WDS) name, mailing address, physical site location	WDS Phone No.	
4. Name and address of responsible agency		
5. Description of Materials	6. Containers No. Type	7. Total Quantity $m^3(yd^3)$
8. Special handling instructions and additional information		
9. OPERATOR'S CERTIFICATION: I hereby declare that the contents of this consignment are fully and accurately described above by proper shipping name and are classified, packed, marked, and labeled, and are in all respects in proper condition for transport by highway according to applicable international and Government regulations.		
Printed/typed name & title	Signature	Month Day Year
Transporter		
10. Transporter 1 (Acknowledgment of receipt of materials)		
Printed/typed name & title Address and telephone no.	Signature	Month Day Year
11. Transporter 1 (Acknowledgment of receipt of materials)		
Printed/typed name & title Address and telephone no.	Signature	Month Day Year
Disposal Site		
12. Discrepancy indication space		
13. Waste disposal site owner or operator: Certification of receipt of asbestos materials covered by this manifest except as noted in item 12		
Printed/typed name & title	Signature	Month Day Year

Objectives: To provide an overview of the requirements and methods for sampling and analyzing asbestos-containing materials before, during, and after an asbestos abatement project.

Learning Tasks: Information in this chapter should enable participants to:

- Become familiar with the various methods for sampling asbestos as a bulk material, airborne fibers, or settled dust.
- Become familiar with the analytical methods used to analyze bulk, air, and settled dust samples.
- Know the common units for reporting airborne fiber concentrations.
- Understand the sampling strategy used for monitoring asbestos abatement projects
- Understand important aspects of final clearance air monitoring including inspection, aggressive monitoring, and clearance criteria.

INTRODUCTION

Sampling and analytical methods are important tools for assessing and monitoring asbestos materials. The applications of sampling and analysis may range from bulk sampling of suspect material to estimating airborne fiber levels before, during, and after an abatement project, to checking surfaces for asbestos-containing settled dust. Collection of reliable data requires a thorough knowledge of the various sampling and analytical techniques that are available and when a particular technique should be used.

This chapter is an introduction to the process of monitoring an asbestos abatement project through the collection and analysis of air samples. After a brief introduction to the many types of sampling methods and various analytical techniques used for asbestos-containing materials, the chapter will focus on items specific to the collection of air samples, including sampling equipment, pump calibration, etc. Next the discussion will focus on the particular advantages and disadvantages of the main alternatives for air sample analysis. Finally, the application of air monitoring and analysis methods to asbestos abatement will be discussed.

SAMPLING AND ANALYSIS – A PERSPECTIVE

Sampling techniques are the first phase in procedures used to collect data representative of the environment. It is analogous to testing a piece of the pie to determine what the entire pie tastes like. If you only taste the crust, then your "sample" will not be representative of the entire pie.

Analytical methods are used to determine what is in the sample. Using the pie again, let's say that an adequate sample was collected (an entire wedge). Let us also use the analytical technique of "touch." This will tell us the size and shape of the sample. From this data, we can estimate the size and shape of the entire pie. "Touch" will also tell us if it is a cream pie. However, this "analytical technique" would not be adequate to determine the flavor or color of the pie.

Just like the pie, the same holds true for sampling and analyzing asbestos. There are many different methods to perform a specific task with each method revealing different bits of information. A person knowledgeable of those techniques selects the appropriate methods to obtain the desired information.

SAMPLING: METHODS AND OVERVIEW AIR SAMPLING

Air sampling involves drawing a known volume of air through a filter and analyzing that filter for the presence of asbestos fibers. The filter is housed in a plastic cassette, which is attached to a sampling pump with flexible tubing. The sampling pump can be either electric (plug in) or battery powered and is calibrated to draw a known volume of air through the filter material over a given period of time (usually expressed in liters of air per minute [lpm]).

Two basic air sampling methodologies are area and personal monitoring. Area samples are taken with a pump, tubing, and filter cassette (called the sampling train), placed at breathing zone height at some stationary location. Personal samples are collected from within the breathing zone (as close to the nose and mouth as possible) of an individual, but outside the respirator. Personal samples are collected in the same manner as area samples, except that the pump is hung from a belt around the worker's waist and the filter cassette is attached, pointing downward, to the worker's lapel or collar.

Specific techniques for the collection of area and personal air samples as they pertain to asbestos abatement projects will be detailed later in this chapter.

OSHA	**29 CFR 1926.1101** Licensed inspector or CIH to determine if TSI or Surfacing non-ACM. Industrial hygienist to determine if other materials are non-ACM
EPA United States Environmental Protectio Agency	**NESHAP 40 CFR 61** No Qualifications for person performing a inspection, but inspection is required to be thorough.
EPA United States Environmental Prote Agency	**AHERA - 40 CFR 763** Requires AHERA certified inspector for inspections and sampling
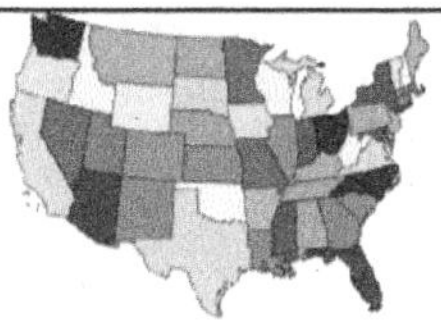	**States** Individual state regulations vary and can be more stringent than Federal requirements. State Regulations may have different trigger numbers regarding quantities and types of asbestos activities.

BULK SAMPLING

Bulk sampling is the technique used to collect samples of suspect materials such as fireproofing, pipe and boiler insulation, and acoustical spray. This sampling is usually conducted during the building inspection/hazard assessment and provides data for decisions on control measures. If bulk sampling data is not available to the contractor during his walkthrough survey, he must choose to have an accredited inspector collect some bulk samples at that time.

A small sample of suspect material is carefully collected and placed in a container (35mm film canisters work well). Anyone taking bulk samples in a school, industrial facility, public or commercial building must be accredited according to U.S. EPA regulations, and should always wear a cartridge respirator and protective clothing when dirty conditions are encountered. Bulk samples are analyzed by an analytical laboratory using Polarized Light Microscopy (PLM) to determine the presence, type, and percentage of asbestos in the sample. Bulk samples can also be analyzed by electron microscopy, which may be useful for analysis of floor tile or other products where asbestos may have been processed into very small and fine fibers.

SETTLED DUST

Note: (Please check with local and state requirements for dust sampling analytical techniques.)

Sometimes, it is beneficial to determine whether the settled dust within a facility contains asbestos. For instance, during the building inspection/survey when investigating for the presence of asbestos-containing materials (ACM), an owner may request that the inspector determine whether asbestos fibers are being released in the building environment. In the past, an air sample might have been collected to determine whether airborne asbestos fibers are present. However, the EPA does not recommend the use solely of air sampling for this purpose as it tends to provide only a "snapshot" picture of building conditions. In the EPA publication, Managing Asbestos in Place (Green Book) (EPA 20T-2003, July 1990), EPA does say that a well-designed air sampling program used in conjunction with comprehensive physical and visual inspections as part of an O&M Program may provide useful information. As an alternative to the use of air sampling, samples of settled dust may be collected to indicate fiber release from ACM.

Sampling settled dust can be accomplished in many ways. Dust can be collected by scraping an area (with a credit card, for instance) and placing the material in a small container for analysis as a "bulk" sample by polarized light or electron microscopy. Alternatively, samples can be collected by "micro vacuuming" an area with a filter in a cassette that is attached to a sampling pump. The filter can be analyzed by polarized light microscopy, or preferably, by electron microscopy. Other techniques for dust sample collection include wipe sampling (where a filter or other material is used to wipe an area) or tape sampling (where cellophane or similar tape is used to collect the dust). The U.S. EPA and the American Society for Testing and Materials (ASTM) are currently developing protocols for settled dust collection and analysis.

It is important to note that most settled dust sampling will typically provide only qualitative results and that any quantitative results must be interpreted with caution. In other words, dust samples should only be used to determine the presence or absence of asbestos fibers in accumulated dust, and not as a tool to determine the amount of asbestos fibers being released from a particular material or the likelihood that these fibers will become airborne again. Also, the absence of asbestos fibers in settled dust does not necessarily mean asbestos fibers are not being released, just that none were present (or detected) in that particular accumulation of dust.

AIR SAMPLING EQUIPMENT

SAMPLING PUMPS

Pumps are the backbone of the air sampling process, providing the means by which air is drawn through the filter that is housed in the cassette. Sampling pumps are typically categorized as either high volume (electric) or low-volume (personal) pumps, which are usually battery powered. High volume pumps are typically larger and heavier than battery powered pumps and cannot pump flow rates that exceed 16 liters of air per Minute (lpm) through the filter. Typically, high volume pumps are used for area air sampling. One advantage of high volume pumps is their ability to draw large volumes of air through the filter in a relatively short period of time. Since being able to detect low

concentrations of airborne asbestos fibers relies, in part, on sampling large volumes of air, high volume pumps are useful for sampling in environments where low levels of airborne asbestos are expected (e.g., outside of a work area containment, or following the cleanup of an abatement project).

Battery powered, or personal sampling pumps are small, lightweight pumps usually encased in a hard plastic shell. These pumps typically draw from one-half to four liters of air per minute through the filter and are ideal for indexing workers' exposure (or potential exposure, when wearing a respirator) to airborne asbestos fibers. Characteristically light weight, these pumps are easily worn on a worker's belt with little discomfort.

FILTERS

Two main types of filter material are used to sample for airborne asbestos fibers. Mixed cellulose ester (MCE) or membrane filters are the most common and have the widest use. MCE filters are cellulose strands bound together in a web called "tortuous pore" and display a very irregular surface when observed under magnification. Polycarbonate filters, on the other hand, are thin sheets of plastic with holes punched in them by neutrons and enlarged by an alkali bath. Once quite popular for applications such as final clearance air monitoring (where analysis was to be performed by electron microscopy), polycarbonate filters should be used with caution and are being used less frequently because of fears of fiber loss from the smooth filter surface during sample handling and transport.

Regardless of the type used, filters are characterized by their diameter (of exposed surface) and their pore size. Table XIV-1 details the recommended filter types, pore size, and diameter for the sampling method and analytical alternative used.

All filters are housed in a sampling cassette, which includes a cap, extension cowl or retainer ring, the filter, a MCE diffuser when collecting TEM samples, a support pad, and a cassette base. Figure XIV-1 details a typical sampling cassette configuration.

PUMP CALIBRATION

As previously mentioned, calculation of air sampling results are dependent, in part, on the total volume of air sampled. The total volume of air sampled is the flow rate of the sampling pump (liters of air per minute or lpm) multiplied by the time (in minutes) that the pump ran. Accurate calibration of the pump flow rate, then, is very important in the calculation of sample results. The EPA and OSHA require that sampling pumps be calibrated before and after each use, and it is required that calibrations be recorded by the person conducting sampling.

Although not always practical, a primary calibration standard is the best way to determine the flow rate of a sampling pump. A primary calibration standard is one that is known to have the highest degree of accuracy and repeatability when determining a pump's flow rate. Typically, a one-liter flow bubble buret or equivalent is used as a primary calibration standard for air sampling pumps. From this, a rotameter, which is smaller and more durable, yet less accurate than the flow bubble buret, can be calibrated and taken into the field to calibrate each sampling pump before and after use.

It is important to ensure that persons performing air monitoring are routinely calibrating their sampling pumps. Regular requests of calibration data or requiring this data to be included in reports of sample results are two ways to help maintain the technical and legal validity of sampling data.

TABLE XIV – 1

RECOMMENDED FILTER – BASED ON SAMPLE TYPE AND ANALYTICAL ALTERNATIVE

PHASE CONTRAST MICROSCOPY	NIOSH 7400	25 mm MCE, 0.45-1.2 µm pore size	25 mm MCE, 045-1µmm pore size*
	OSHA reference Method (ORM)	25mm or 37mm** MCE, 0.8-1.2 µm pore size	
TRANSMISSION ELECTRON MICROSCOPY	AHERA Mandatory Method		25 or 37mm MCE 0.45µm pore size 25 or 37 PC, 0.4 µm pore size
	Yamate		37mm PC, 0.4 µm pore size
	NIOSH 7402		25 mm MCE, 0.8, 1.2 µm pore size
	Burdett & Rood		25mm MCE, 0.1, 0.8 1.2 µm pore size 25 mm PC, 0.45, 0.8 µm pore size***

MCE= mixed cellulose ester filter

PC= polycarbonate filter

* per 40 CFR, Part 763, Subpart E, Appedix A (AHERA)

** Use of 37mm cassette in ORM requires written justification.

*** Burdett, G.J. "Proposed Analytical Method for Determination of Asbestos in Air"

Figure XIV-1

TYPICAL TEM SAMPLING CASSETTE CONFIGURATION

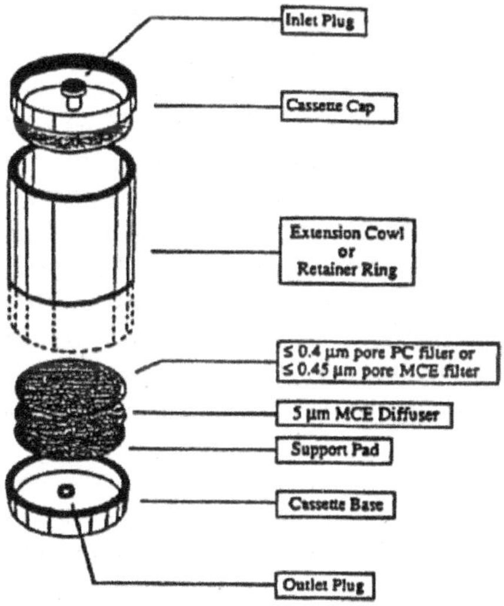

ANALYTICAL ALTERNATIVES

The primary analytical techniques used for analyzing airborne fibers collected on a filter are: Phase Contrast Microscopy (PCM), Transmission Electron Microscopy (TEM), and Scanning Electron Microscopy (SEM). Table XIV-2 summarizes the advantages and disadvantages of each. The Fibrous Aerosol Monitor (FAM) is an instrument that can be used in the field to obtain an index of airborne fiber levels. Applications of each of these methods (PCM, TEM, SEM, and FAM) in the analysis of air samples for asbestos are discussed in this chapter.

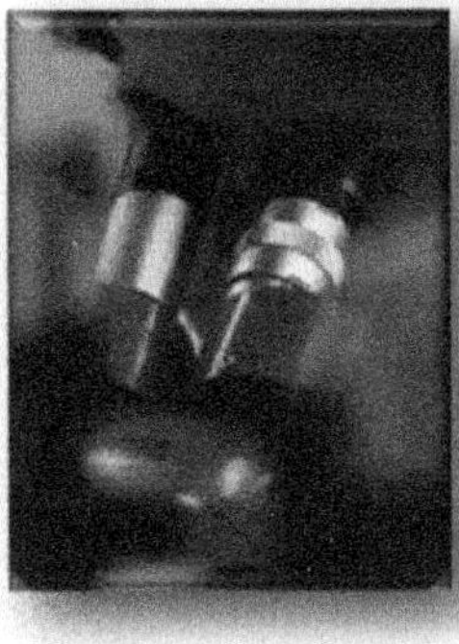

PHASE CONTRAST MICROSCOPY

Phase Contrast Microscopy (PCM) is a technique using a light microscope equipped to provide enhanced contrast between the fibers collected and the background filter material. Samples for analysis by PCM are collected on either a 25 mm or a 37 mm mixed cellulose ester (MCE) filter with a 0.45 to 1.2 micrometer pore size. Filters are then prepared by either a liquid chemical solution or an acetone vapor that renders the filter material optically transparent. The filter is then examined under a positive phase contrast microscope at a magnification of approximately 400x. Fibers are sized and counted using a calibrated reticle fitting into the microscope eyepiece. PCM samples is inexpensive (check with your instructor) and can be performed on the job site in a few hours.

Phase contrast microscopy is frequently referred to as the light microscopy method, the filter membrane method, or the NIOSH method. PCM is the analytical method specified in the Occupational Safety and Health Administration (OSHA) Asbestos Standards. PCM was first used to monitor asbestos workers' exposure in asbestos product manufacturing or milling operations for prevention of asbestosis. This method does not identify the fibers it counts and only counts those fibers longer than 5 micrometers and wider than about 0.25 micrometers. Because of these limitations, analysis by PCM typically provides only an index of total concentration of airborne fibers in the environment monitored. As the proportion of the airborne fibers that are less than 0.25 micrometers in diameter increases (i.e., non-industrial settings such as asbestos abatement projects), PCM becomes a less reliable analytical tool.

There are two primary fiber-counting methods for phase contrast microscopy. The NIOSH 7400 method is a sample collection and analysis procedure that has been refined over the years to provide fiber counts with a reliable limit of detection and which are reproducible. The OSHA reference method (ORM) is specified in the OSHA Asbestos Standards and contains modifications to the procedures outlined in the NIOSH 7400 method.

TRANSMISSION ELECTRON MICROSCOPY (TEM)

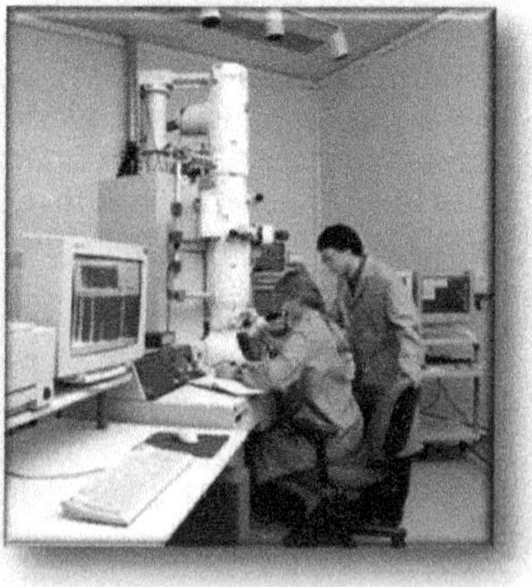

Transmission Electron Microscopy (TEM) is a technique that focuses an electron beam onto a thin sample mountedin the microscopy column (under a vacuum). As the beam transmits through the sample, an image resulting from the varying density of the sample is projected onto a fluorescent screen. Air samples for TEM analysis can be collected on either mixed cellulose ester or polycarbonate filters and are prepared using direct transfer techniques (per EPA regulations). Direct preparation allows for the transfer of a carbon-coated replica of the filter material (with embedded fibers and particulates, etc.) right on to a copper grid suitable for TEM analysis. Indirect transfer techniques require an intermediate step that may break up fiber bundles, resulting in an increased fiber count.

Several methods exist for the preparation and analysis of air samples by electron microscopy. Most significant are the mandatory and non-mandatory TEM methods set forth as appendices to 40 CFR 763, Subpart E (AHERA Regulations). These methods are to be used, with restrictions, for analysis of final clearance air samples on school abatement projects. Other methods include the NIOSH 7402, the Yamate Method, and the Burdett and Rood Method. Depending on the method used, preparation of the sample can take as much as 24 hours (or more) and analysis can take several hours to a day or more. Also, limited availability of labs capable of providing TEM analysis often leads to turnaround times on the order of days or weeks. As the number of laboratories with TEM capability continues to increase, the cost and turnaround time will go down. Costs for TEM analysis will vary per sample. Check with instructor!

SCANNING ELECTRON MICROSCOPY (SEM)

Scanning Electron Microscopy (SEM) is a technique that uses a finely focused electron beam on the sample surface to generate an image of the surface topography. A magnified image is produced on a viewing screen. Air samples for SEM filter counting are collected on a polycarbonate filter with a 0.45 micrometer pore size. The cost is about $150-$300 per sample and may require several days to obtain results.

SEM can identify large fibers by morphology (physical appearance) and elemental analyses when equipped with an energy dispersive x-ray analysis system. Fibers that are 0.05 micrometers in diameter are about the smallest that can be detected using SEM under optimal conditions. This method has fiber identification problems with thin fibers and flat, platy particles that display poor contrast. Also, there is no standard protocol for this method. Currently, SEM provides somewhat better information than PCM analysis, but the method cannot be used to conclusively identify or quantify asbestos.

FIBROUS AEROSOL MONITOR (FAM)

The fibrous aerosol monitor is an instrument that uses laser light and electrical field technologies to instantaneously analyze the fiber content of the air. The instrument provides a continuous measurement, with direct readout of the number or concentration of airborne fibers. The FAM can be used in conjunction with a strip chart recorder to provide a record of air quality conditions. Typically used as a barometer of airborne fiber levels rather than a precision testing device, the FAM's more useful function is to alert personnel to any sudden elevation of the area fiber count. If the FAM is used on a project, it should be used in conjunction with other traditional air sampling techniques and not in place of them.

This instrument does not distinguish fiber types and cannot discriminate between fibers and certain particles that have sufficient shape irregularities to possess fiber characteristics. The FAM does not detect fibers less than approximately 0.2 micrometers in diameter. Laboratory tests indicate that FAM concentration readings are generally within ±5 percent of the optical membrane filter (PCM) count.

NOTE: (Please note that some states do not authorize either SEM or FAM analytical methods for asbestos analysis.)

Figure XIV -2 is a sampling log form included in appendix A-IIA (Mandatory TEM Method) of 40 CFR part 763, Subpart E (AHERA)

Figure XIV-3 is a sample TEM report format also included in the Mandatory TEM Method.

FIGURE XIV 2
SAMPLING LOG FORM

Sample Number	Location of Sample	Pump. I.D.	Start Time	Middle Time	End time	Flow Rate

FIGURE XIV -3
EXAMPLE LABORATORY

Laboratory ID	Client ID	[----------- FILTER MEDIA DATA ----------] Type	Diameter mm	Effective Area, mm2	Pore Size, μm	Analyzed Area, mm^2	Sample Volume, cc

INDIVIDUAL ANALYTICAL RESULTS CONCENTRATION

Laboratory I.D.	Client I.D.	# Asbestos Structures	Analytical Sensitivity, s/cc	Structures/mm^2	Structures/cc

The analysis was carried out to the approved TEM method. This laboratory is in compliance with the quality specified by the method.

TABLE XIV-2

COMPARISON OF AIR SAMPLE ANALYSIS ALTERNATIVES

	PCM	SEM	TEM
Standard Methods	NIOSH 7400 Method	No standard Method	AHERA Method And others
Quality Assurance Materials	Proficiency Analytical Testing Program; no NIST reference materials	No lab testing, Or NIST reference materials	Limited lab testing NIST reference available
Cost	Check with instructor	Check with instructor	Check with instructor
Availability	Most available	Least available	Less available
Time Requirements	1 hr preparation & analysis, 6 hrs. turnaround	4 hr preparation & analysis, 6-24 hrs. turnaround	4-24 hr preparation & analysis, 2-7 days turnaround
Sensitivity (Thinnest Fiber Visible)	0.15 μm at best; 0.25 μm typical	0.05 μm at best; 0.20 μm typical.	0.0002 μm at best 0.0025 μm typical.
Specificity	Not specific for asbestos.	More specific than PCM but not definitive for asbestos	Definitive for asbestos (Level III – TEM with EDXA & SAED)
Collection Filters	0.45-1.2 μm cellulose ester.	0.4-0.8 μm poly-carbonate best, cellulose ester also used.	0.4 μm polycar-bonate, or 0.45 μm cellulose ester if organic contaminants present.

(μm = micrometer)

SAMPLING STRATEGIES AND PROCEDURES FOR AN ABATEMENT PROJECT

Monitoring of an asbestos abatement project through the use of air sampling is a complex task involving both personal and area samples. Air samples can be taken under such widely varying condition that no two sample results are alike, even in the same containment area on the same day. The following chapter outlines some of the fundamental ways air sampling is used to monitor asbestos abatement projects.

NOTE: (Please see local and state requirements for area air sampling.)

AIR SAMPLING BEFORE ABATEMENT

BEGINS Prevalent Level Samples

Area air sampling conducted before abatement activities begin to estimate the existing airborne fiber concentrations inside and outside the building is termed prevalent level sampling. This type of sampling is also referred to as "pre- abatement" or "background." Results can be used as control data for comparing sample concentrations detected during the abatement project. Prevalent level sampling provides good data for documentation purposes. It is particularly useful when an abatement project is conducted in a portion of a building, with other areas of the building remaining occupied. Airborne fiber levels monitored in these occupied areas during the abatement project should never exceed the indicated prevalent level in these areas before the project began.

Because low airborne fiber concentrations are typically found prior to abatement activities, a large volume of air should be sampled to obtain a low detection limit. Simply stated, detection limit is the lowest value that can be reliably reported for the sampling and analytical methods used. The volume of air measured to obtain a 0.01 fiber per cubic centimeter of air (fiber/cc)[1] detection limit should range between 3000 to 4000 liters, depending on the filter size and counting method used. Samples can be collected at a flow rate typically in the range of 2 to 10 liters per minute.

Prevalent level samples should be collected throughout the building as well as in the areas where abatement will take place. As a rule of thumb, one sample should be taken for every 50,000 cubic feet (5,000 sq. ft. with 10 ft. ceilings) of building space (minimum of 3 samples per building). At least two samples should be collected from outside the building. Because results of prevalent level sampling are used as baseline data, the same sampling and analytical techniques should be used for pre-abatement samples as well as subsequent samples (i.e., indoor/outdoor removal project, clearance samples, etc.). Since most samples taken during an abatement action are analyzed by phase contrast microscopy (PCM), most prevalent level air samples should be analyzed by PCM as well. Occasionally, the building owner or manager will want more "absolute" data for baseline purposes.In this case, additional prevalent level samples should be collected and analyzed by Transmission Electron Microscopy (TEM). One should not attempt to compare prevalent level samples analyzed by TEM with abatement samples analyzed by PCM as no sound conversion exists between these two analytical methodologies.

Another note of caution about prevalent level air samples involves comparison with air samples taken to aid in determination of the completeness of a response action (final clearance). Although the U.S. EPA does allow for comparison of samples taken both inside and outside the removal project, these samples must be taken at the same time to ensure identical environmental conditions. Outside prevalent level air samples taken well before final clearance samples cannot give an accurate indication of ambient conditions at the time of final clearance.

AIR SAMPLING DURING AND AFTER THE ASBESTOS ABATEMENT PROJECT

Personal Sampling

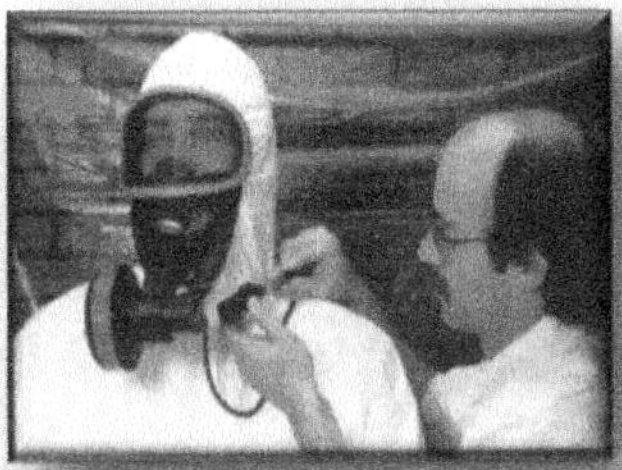

Personal sampling is conducted during a renovation or abatement project to determine employees' exposure (outside the respirator) to airborne fibers. Data from personal monitoring serves many purposes. Monitoring during an abatement project is required by the OSHA Asbestos Standards (29 CFR 1910.1001 and 1926.1101). Under OSHA and hazard communication laws, employees have the right to know the asbestos concentrations to which they are exposed and what measures are being taken to protect them. Also, results of personal sampling can be used to select proper respiratory protection for an employee if conditions warrant something other than Type C respirators (see Respiratory Protection Chapter). Data from personal monitoring can be used as an indication of effective removal or control techniques that result in the lowest employee exposure. This, in turn, reduces the potential of asbestos-related diseases and the risk to the worker.

Personal samples should be collected during the first full day of removal activity. It is generally accepted that this "initial" monitoring must be performed on at least 25% of the work force involved in the project and that it is required regardless of the respiratory protection used (no waiver for Type C respirators). Initial monitoring must be conducted to determine both the 8-hour time weighted average **and** the 30-minute short term excursion exposure.

Periodic monitoring must be performed when the type of material being removed or the location of removal changes. For the OSHA Construction Industry Standard (29 CFR 1926.1101), it is generally accepted that this means daily air monitoring on 25% of the work force is required. Here again, monitoring for both the 8-hour and 30-minute exposures must be conducted. The exception to this requirement involves workers using supplied air, positive pressure (Type C) respirators. Also, when monitoring indicates exposures below the PEL (0.1 f/cc) and/or the excursion limit (1.0 f/cc), daily monitoring may be terminated until conditions significantly change. However, it is prudent to conduct personal sampling on a periodic basis regardless of the type of respiratory protection used.

Personal samples are typically collected at a flow rate of 0.5-2 liters per minute (lpm). Samples for asbestos exposure should be taken to determine the 8-hour, time-weighted concentration, as well as the 30-minute excursion limit. Over an eight-hour period, filters may have to be changed several times to prevent overloading. Results of each sample are put into the following equation to obtain a time-weighted average concentration for the total sampling period:

$$\frac{C_1T_1 + C_2T_2 + C_3T_3 + \ldots}{T_1 + T_2 + T_3 + \ldots} = \text{Time Weighted Average,}$$

where C is the fiber concentration expressed as f/cc and T is the duration of sample in minutes.

Table XIV-3 summarizes the primary requirements for personal exposure monitoring as specified in the OSHA Asbestos Standard for the construction industry (29 CFR 1926.1101).

TABLE XIV – 3
SUMMARY OF OSHA (29 CFR 1926.1101) EXPOSURE MONITORING REQUIREMENTS

SAMPLING

I. Personal monitoring for 8 hour TWA, representing full shift exposure, and/or 30 minute short term exposure.

A. Initial Exposure Assessment (25% of workforce recommended) at the start of each job.
   1. Based on exposure monitoring, observations, information, and calculations.

B. Negative Exposure Assessment
   1. Demonstrate exposure below PEL (0.1 f/cc) and/or excursion limit (1.0 f/cc)
   2. Demonstrate a negative exposure assessment through objective data
   3. Demonstrate a negative exposure assessment through data collected within the previous 12 months indicating exposures below the PEL./EL
   4. Demonstrate negative exposure assessment through initial monitoring indicating exposures below the PEL/EL

C. Periodic monitoring (25% of workforce recommended) conducted daily

1. Waiver for the use of supplied air (type-C) respirators operated in positive pressure mode

2. Termination of periodic monitoring if statistically reliable sampling methods indicate exposures below PEL and/or excursion limit

II. Method of monitoring and analysis as outlined in Appendix A of standard (OSHA Reference Method)

III. Employees must be notified of monitoring results as soon as they are available.

A. Must be in writing

IV. Employees or their representative must be allowed to observe monitoring

ANALYSIS

I. Analysis by Phase Contrast Microscopy (PCM) using OSHA Reference Method (ORM)

- Magnification at 400X
- Walton Beckett eyepiece graticle
- Scope calibrated with HSE phase shift test slide
- Blank analysis required
- Filters cleared with acetone/triacetin method or equivalent
- Count fibers greater than five micrometers in length with a length to width aspect ratio of 3:1 or greater
- Follow counting rules in ORM

II. Quality control procedures

- Intralaboratory
- Interlaboratory
- All analysts must successfully complete NIOSH 582 course or equivalent

Area Air Sampling Inside the Work Area

In addition to personal samples, area air samples are collected inside the work area daily to determine the concentrations of airborne asbestos fibers. Two to three samples of 60 to 120 liters of air are usually adequate to index the airborne fiber concentrations inside the work area. The analytical data from these samples can be used on a relative basis to monitor work conditions from one day to the next. A radical increase in area concentrations would signal that work practices needed to be adjusted.

Area air samples are typically collected in different places within the removal area from one day to the next. Typical sample locations include: the immediate vicinity of removal (i.e., on scaffolding used to access ACM), "upwind" or "downwind" from removal (in terms of airflow created by pressure differential from exhaust units - see chapter on Confining and Minimizing), in the dirty equipment room or shower area of the decontamination unit, or near "critical" barriers.

Inside area air samples taken during the removal project are typically collected using a portable, battery-powered "personal" sampling pump because they are easy to transport and clean and because sampling large volumes of air is not necessary since these samples are used as an index of day to day condition. Analysis should be performed using phase contrast microscopy.

Area Air Sampling Outside the Work Area/Inside the Building

During the abatement project, samples are collected from locations outside the work area, but inside the building to determine how well asbestos fibers are being contained in the worksite. These samples are especially important in situations where unprotected people are occupying other areas of the building. Potential leakage points where sampling should be conducted include the clean side of the containment barriers separating the work area from occupied parts of the building and inside the clean room of the decontamination unit. If the configuration of the work area is such that sealing the area is difficult due to interference from piping systems running through the area, it is especially critical to sample outside the work area. If the abatement project is being conducted in a multistory building, area air samples should be collected from floors above and below the abatement activity.

A large air volume of 3000 to 4000 liters is necessary to obtain the desired detection limit of 0.01 fibers per cubic centimeter for these samples. High volume pumps can be used to shorten the sampling time so that problems that develop can be detected relatively quickly. Phase contrast microscopy is generally the analytical method used for these air samples.

Area Air Sampling Outside the Building

Area air samples are placed in locations outside the building during the abatement project to detect leakage of fibers from the worksite. Typically, pumps are placed at the entrance of the decontamination unit, at doors or windows, near the exhaust of the negative air filtration units, and at the waste load out area. Care must be exercised to ensure that outside samples are not overloaded with dust or other debris. Also, filter cassettes placed too close to the exhaust stream of the negative air filtration units will not give a meaningful indication of fiber leakage from these units. Samples should either be collected isokinetically, which demands special sampling equipment and a greater degree of expertise, or placed downstream from the exhaust unit where fibers are more likely to be captured by the sampling apparatus.

Air Sampling After Final Cleanup of Work Area

Area air sampling is conducted upon conclusion of an asbestos abatement project to estimate the airborne concentration of residual fibers. The area must pass a thorough visual inspection for remaining material before final clearance sampling is initiated.

A visual inspection process is typically conducted in two phases. First, an inspector determines the completeness of removal. If any visual ACM is clinging to the substrate, then the removal is not complete and the inspection does not continue. Once the removal is determined to be complete, the work area is inspected for cleanliness. If any dust or debris is found, the area is recleaned before air sampling commences.

Final clearance air samples are typically collected using high volume pumps to draw a predetermined volume of air. The number of samples collected depends on the amount of ACM affected by the response action, where the project is taking place, and the sampling and analytical procedures being followed. Samples should be collected using aggressive techniques. Aggressive air sampling involves physically or mechanically agitating the air in the work area during the sampling process. Typically, a one horsepower leaf blower is used on all surfaces in the work area to dislodge any residual fibers. Then, a standard box fan is left on for the duration of sampling to keep any dislodged fibers airborne. The purpose of final clearance air sampling using aggressive techniques is to produce a "worst case" scenario. If the work area passes the final clearance level in this "worst case" environment, then the likelihood of airborne asbestos fiber levels ever rising above the clearance level is remote.

Leaf Blower used for Final Clearance

Clearance samples are typically analyzed by phase contrast microscopy or transmission electron microscopy. Ideally, PCM and TEM are used in combination as a two-stage process for final clearance sampling. Phase contrast analyses can be used to determine if any gross contamination remains in the work area. If the PCM samples indicate airborne fiber levels are below 0.01 f/cc using aggressive sampling techniques, then the samples are submitted (or new samples are collected) for analysis by transmission electron microscopy. As discussed earlier, TEM is the analytical method recognized as having the best resolution and positive fiber identification capabilities, and is required in most cases when performing clearance sampling under the AHERA regulations.

Table XIV-4 outlines the final clearance air sampling, analytical sequence and clearance level requirements for asbestos abatement projects conducted in school buildings per the Asbestos Hazard Emergency Response Act (AHERA) regulations (40 CFR part 763, subpart E).

TABLE XIV - 4
AHERA FINAL CLEARANCE
TESTING SAMPLING

I – Sampling Agency

- ***Written quality control documents, verify compliance***
- ***Completely independent of abatement contractor***

II - Sampling Equipment

- *Commercially available cassettes*
- *Prescreen loaded cassettes for background levels*
- *Filter and cassette requirements per Table XIV -1 & Figure XIV-1*
- *Reloading of cassettes not permitted*
- *High Volume sampling pump*

III – Sampling

- Visual inspection prior to sampling
- Final ("critical") workplace barriers to remain in place
- Calibrate pumps before and after each use
- Pump flowrate of 1-10 liters per minute (1pm) for 25 mm cassettes (proportionally higher for larger diameter filters)
- Clearly label all samples
- Maintain log of all pertinent sampling data
- Perform leak check on sampling train
- Isolate pump vibration from filter cassette
- Orient cassette 45 ^{o} downward from horizontal
- Use aggressive sampling techniques
- Collect a minimum of 13 samples if TEM is used: 5 samples if PCM is to be used:
  5 per abatement area
  5 per ambient area (outside the abatement area) (TEM only)
  2 field blanks (required for TEM, recommended for PCM)
    1 near entrance to work area
    1 at ambient site
    1 sealed blank (TEM only)
- Turn Sample cassette upright before turning pump off

ANALYSIS

I – Transmission electron Microscopy

- Required for projects involving ACM in amounts greater than 160 square feet or 260 linear feet.

II - TEM Analysis Sequencing

- Collect at least 13 samples
- Analyze five inside samples
- If greater than or equal to 1,199 liters of air collected (for 25 mm cassettes) Or 2,799 liters of air (for 37mm cassettes), area passes if arithmetic mean less than or equal to 70 structures per square millimeter of filter area (70) s/mm^2)
- If less than 1,199 liters (2,799 of air is collected, or if the arithmetic mean is greater than 70/s/mm^2, analyze three blanks
- If arithmetic mean of blanks is greater than 70 s/mm^2, terminate analysis, identify source of contamination, collect new samples
- If arithmetic mean of blanks less than 70 s/mm^2, analyze outside samples and compare using Z-test
- If Z-statistic is less than or equal to 1.65 response action compete
- If Z-statistic is greater than 1.65 reclean and resample

III - Phase Contrast Miscroscopy

- Allow if amount of ACM involved in the abatement project is less than or equal to 160 square feet or 260 linear feet

Phase Contrast Microscope Configuration

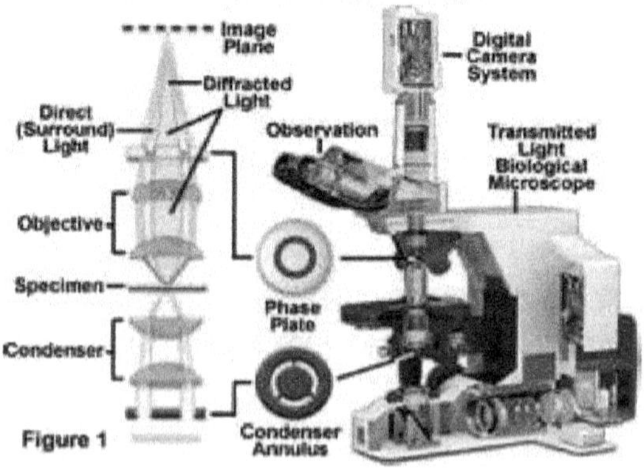

Figure 1

Objectives: To discuss post removal lockdown operations, available substitute materials, and practices use for replacing the original asbestos materials.

Learning Tasks: Information in this chapter should enable participants to:

- Become familiar with the "lockdown" operation and how it fits into the final cleaning stages of a project.
- Understand procedures used for applying lockdown materials to the substrate after removal of asbestos- containing materials has taken place.
- Become familiar with the criteria used in selecting certain types of materials as effective lockdown agents.
- Understand various factors that may influence asbestos replacement or re- insulation operations.
- Gain awareness of some of the materials currently being used as asbestos-free substitutes and replacements.
- Understand exactly where lockdown and asbestos replacement operations fit into the post abatement sequence of events.

DEFINITION OF LOCKDOWN PROCEDURE

Lockdown is the procedure of applying a protective coating or sealant to a surface from which asbestos-containing material has been removed. Its primary function is to control and minimize the amount of airborne asbestos fiber generation that might result from any residual asbestos-containing debris on the substrate; its purpose is not to cover a poorly performed removal project. Though the substrate may appear to be clean, very small amounts of asbestos- containing material may have become lodged in cracks or crevices that were inaccessible during removal operations.

(Note: Most states do not allow lockdown on abated surfaces until satisfactory clearance air samples have been obtained. Please see state rules and regulations for specific information.)

Lockdown Reasoning

Every asbestos removal project ultimately involves the stripping away of some asbestos- containing material from a permanent substrate or subsurface. Depending upon the surface structure of this substrate or the cohesive strength of the asbestos-containing material to the substrate, there will always be some residual fibers left behind after gross removal has taken place.

It is essential to determine what the substrate is made of before any asbestos is removed. Some of the most common materials found as substrates in buildings include structural steel, miscellaneous metals, cement, corrugated sheet metal, wire mesh, metal piping, plaster "brown coat," and wood. These materials each have different characteristics pertaining to surface structure and cohesive strength.

For example, cement substrates are often porous and pitted (many small grooves on the surface), and steel elements may be rusted and pitted. This uneven type of surface is extremely difficult to clean for two reasons. First, the pits in the surface may have become filled with asbestos-containing material when it was originally sprayed or troweled on the surface and then not removed by wet and scrape methods. Secondly, when the material is scraped away during removal, asbestos-containing materials will be packed tightly into these grooves or pits. Most of the material can then be removed through tedious brushing; however, some fibers will remain. For this reason, it is necessary to develop and follow a lockdown strategy which will effectively control the future release of airborne fibers from porous or uneven surfaces from which removal has already taken place.

LOCKDOWN METHODS

Under the U.S. EPA NESHAP regulations, it is required that all asbestos removal projects be performed using wet methods. In certain instances (i.e., extreme electrical hazards), it may be necessary to greatly limit the use of water since the combination of water and electricity would increase the risk of electrocution. In special cases such as this, other engineering controls should be implemented.

"Wet methods" involve wetting the material to be removed via an airless sprayer with amended water. This wetting will keep the amount of airborne fiber generation minimal and it will also facilitate easier removal of the asbestos- containing material since it will be more pliable. In their "Purple Book" (Guidance for Controlling Asbestos-Containing Materials in Buildings, EPA 560/5-85-024, the EPA recommends using an airless sprayer for applying amended water and encapsulating solutions to minimize airborne asbestos fiber generation.

The recommended method for brushing or cleansing a substrate after gross removal has taken place is to use a <u>nylon</u> brush. This will aid in getting to fibers that may have become lodged in grooves or crevices in the substrate surface. Wetting of the substrate should also take place while this brushing is being performed since the chance of airborne fiber generation is still present. Use of wire brushes is discouraged since wire tends to break down larger asbestos fibers or fiber bundles into fibrils of minute size which are easily dispersed throughout the surrounding air (heavy dispersion can make final cleaning very difficult). In either case, brushing will generate airborne fibers to some degree. Once this brushing is completed, a final cleaning of the substrate should take place in order to ensure that all loose residual fibers are eliminated. It may be necessary to wipe some surfaces with a lint-free rag and dusting agent once it has dried.

After the substrate has completely dried and <u>ONLY AFTER</u> it has passes a thorough inspection for visible residual contamination, application of the lockdown material can begin.

The substrate should be completely cleaned of all visible debris before applying the lockdown sealant. Using double layers of polyethylene on the walls and floors during preparation of the work area will produce greater efficiency at this stage since the inside layer can be removed after the initial lockdown application takes place. Workers performing the lockdown operation should wear disposable protective clothing and respirators suitable for asbestos <u>and</u> organic vapors (if applicable) because the area is still considered contaminated. Respirators which provide protection against both asbestos fibers and organic vapors will be needed if the lockdown material contains volatile organics in liquid form when being applied. However, many lockdown materials currently available are water- based and are considered to present very little respiratory hazard.

When considering a lockdown operation, it is important to follow a logical sequence of events. One good recommended sequence to follow when conducting lockdown operations is outlined below:

1. Complete removal of asbestos-containing material from the substrate.

2. Collect the asbestos-containing waste material and transport it out of the work area enclosure according to appropriate asbestos waste handing procedures.

3. Clean all visible debris by HEPA vacuuming and wet wiping.

4. Conduct a visual inspection of the work area enclosure for any remaining visible debris and reclean if necessary.

5. Only after the work area has passed a visual inspection for asbestos-containing debris and the asbestos waste containers are completely removed from the enclosure, one heavy coat of lockdown sealant should be spray-applied to the substrate in accordance with manufacturer's instructions. At the same time, the inside layer of polyethylene should be misted with a coat of lockdown material. If possible, a test area of suitable size should be sprayed first to observe conditions.

6. After the lockdown material has dried (time is dependent on type of material and the manufacturer's recommendations), the inside layer of polyethylene on the walls and floor of the enclosure should be taken up, treated as asbestos-containing waste, and transported out of the work area.

7. At this point, with the final (outside) layer of polyethylene still in place, a second, more comprehensive visual inspection should be conducted to locate any asbestos materials that may have penetrated to the outside layer and show visible accumulations of dust on surfaces. If asbestos contamination has penetrated to this outside layer, additional HEPA vacuuming and wet-wiping of all surfaces in the enclosure will take place at this time. If desired, a second lockdown misting can be performed on the inside of the final layer to seal in any residual asbestos fibers.

8. If additional lockdown has been performed on the final polyethylene layer of the enclosure, sufficient drying time should be allowed before proceeding. At this point, pre-clearance monitoring may be conducted to determine if any airborne fibers are remaining. In the event that these air sample results indicate the work area is still contaminated, asbestos debris would not be spread throughout surrounding areas. The final poly layer may be taken down and disposed of as asbestos-containing waste when preliminary air sample results indicate an acceptable clearance level.

9. While the special items such as doors, windows, and HVAC systems remain sealed (critical barriers), this is the best time to conduct final clearance air monitoring, using aggressive sampling techniques.

A variety of products can be used for locking down the substrate. These lockdown products are usually applied as sprayed-on liquid type sealants (alternatives for certain situations are latex paint, encapsulating solutions, and concrete sealant). Lockdown material should be applied using an airless sprayer. Additionally, it is important that the lockdown material be compatible with the substrate. Thus, it is important to ensure that cohesion occurs between the two surfaces (substrate and lockdown material), and in some cases, three surfaces (substrate, lockdown, and replacement or re-insulation). For example, latex paint may work well in locking down a porous concrete surface, whereas it would not be acceptable for use on metal piping since it would peel and crack. Caution should also be used so that the lock-down material does not present an additional hazard during application and anticipated use/conditions. Contractors should obtain all available information on the substance (i.e., toxicity, volatility, fire ratings).

Material Safety Data Sheets (MSDS) are a good source of information. These sheets should be available from both the manufacturer and distributor of the material. Under OSHA requirements for Hazard Communication, contractors are required to train their employees so that they understand and know how to use the data sheets for any potentially hazardous product. It may be necessary to request additional data from other sources on the fire ratings of some materials. All information should be obtained and evaluated prior to beginning the project. Many encapsulating or lockdown materials are not fire rated (they may actually be more flammable) and thus could greatly reduce or eliminate the fire rating of some replacement materials.

A good, recommended practice is to use color tinting when applying multiple coats of lockdown materials. This will make it easier to visually check that all areas of the substrate have been covered with the lockdown substance. One coat of lockdown substance will usually be adequate to prevent the generation of airborne residual fibers. In some cases, additional coats, of a different color, may be needed for cosmetic purposes. Also, if the lockdown material is being applied to metal substrate surfaces, it may be most advantageous to apply one fairly heavy coat of primer to act as both a lockdown material and corrosion inhibitor, regardless of whether or not re-insulation is taking place.

REPLACEMENT PRODUCTS ASBESTOS SUBSTITUTES

Once the lockdown sealant has been applied and after final clearance monitoring has taken place, often the next step is to reapply an adequate substitute for the asbestos-containing material that was originally present. In most cases, the original asbestos-containing material was probably used as fireproofing, thermal insulation, condensation control, or acoustical insulation, or possibly just for decoration. Therefore it is imperative that the substitute material be capable of the same functions and has similar properties relative to the original asbestos-containing material.

This substitute material should be investigated and chosen during the planning stage of the project. Local building code requirements should also be checked prior to material selection.

SPRAY ON INSULATION

The architect or engineer, as a member of the project team, should have the capacity to investigate and recommend various types of re-insulating materials or asbestos substitutes. This person should be most familiar with the chemical and physical properties ofthe various materials available. He/she should also be familiar with the building structure; specifically, the acoustics and fire ratings. Additionally, the industrial hygienist, who is also a member of the project team, may be able to evaluate the substitute or re-insulation material for potential health hazards or other problems. Working in conjunction with one another, these two members of the project team should be able to decide on an adequate asbestos replacement material while the job specifications are being drawn up. The actual design of the asbestos replacement operation should be conducted by an architect with experience in the field, and licensed to practice in the state where the project is located. Individuals from other disciplines who are not trained in building design do not have the proper background or legal standing to conduct such design work.

In many instances, the non-asbestos-containing substitute will be applied to a substrate as a sprayed-on coating. Various materials are available depending on what the specific purpose will be. There are literally hundreds of commercially available asbestos-free replacement or re-insulation materials. There are both naturally occurring and manmade products that have been found to be useful substitutes for asbestos. Some of these materials include the following;

- Vermiculite*
- Perlite
- Fibrous Glass
- Mineral Wool
- Rock Wool
- Steel Wool
- Treated Cellulose
- Calcium Silicates
- Various Metals
- Carbon Fibers
- Aramid Fibers
- Foamed Rubber
- Processed Cork
- Ceramic Fibers
- Synthetic Fibers

* is assumed to contain asbestos

These are only a few of the many products that have been used or are being experimented with as substitutes for asbestos. Each of these materials has some special quality that may make it a suitable replacement material for asbestos in certain cases. The difficulty lies in the fact that none of these materials alone can be as universally applied for as many different functions as asbestos was. Also, the costs associated with manufacturing some of these substitutes tend to be prohibitive. As research continues, however, the costs of many of these products have been to decrease. As an added precaution, bulk samples of replacement materials should be analyzed to verify non-asbestos composition.

Finally, it is important to ensure that the materials chosen for lockdown and replacement or reinsulation are compatible. It may be necessary to perform small "sample" applications in inaccessible or non-visible areas of a structure to determine if the materials will stick to each other and also to check the appearance of the materials after application.

The practice of locking down the substrate has rapidly gained acceptance as a critical step in the cleaning sequence of asbestos removal procedures. It can aid in minimizing the regeneration of asbestos fibers left on the substrate after removal, which will in turn help to make the area acceptable for re-occupancy more quickly. Additionally, asbestos abatement contractors are often called on to perform reinsulation operations once they have removed asbestos-containing materials from a facility. Therefore, it is essential that they have a thorough understanding of how and why asbestos was used in certain cases and which products may serve as the best substitutes. Many asbestos abatement contractors will sub-contract the reinsulation work to another contractor after they have finished the asbestos abatement project. Even if this is the case, it is important for asbestos abatement contractors to be aware of general procedures used.

Because some residual asbestos contamination may remain and be sealed by the lockdown process, later gross demolition of the substrate could release these fibers. It is therefore best to document all abatement work and to install caution labels under certain conditions. Operations and Maintenance Plans should also be modified accordingly following abatement project.

The following is a list of products and materials that have found use as substitutes for asbestos containing materials. Some of these products are no longer manufactured, and some of the companies listed may no longer exist.

Note: Exposure to some of the materials listed may cause health problems associated with inhalation of the material. All cautions should be noted, and proper safety and health measures should be taken in accordance with the manufacturer's instructions.

SUBSTITUTE MATERIALS FOR ASBESTOS-CONTAINING MATERIALS

I. Pipe Insulation
II. Sprayed-On Materials
III. Wallboards/Panels
IV. Fabric/Textiles
V. Cements/Plastics
VI. Brake Lining/Disc Pads
VII. Packing/Filler Materials
VIII. Gasket Materials

1. Substitute Materials for Pipe Insulation

TRADE NAME	MANUFACTURER	SUBSTITUTE MATERIAL
Aerotube	Johns-Manville	Foamed Plastic
Alpha-Mritex Style 1925	Alpha Associates	Glass Fiber
Alpha-Maritex #3111-RW	Alpha Associates	Fiber Glass
Armaflex 22	Armstrong Cork	Foamed Plastic
CPR	Upjohn	Plastic
Crown	Fibreglas, Ltd.	Fiber Glass
Fit-Rite	Fibrous Glass Products, Inc	Glass Fiber
Flame-Safe	Johns-Manville	Glass Fiber
Glo-Brite	Glo-Brite Products	Polyfoam
GPC	Johns-Manville	Glass Fiber
Hewflex	H.E. Werner, Inc.	Polyurethane Foam
Kaowool	Babcock & Wilcox	Ceramic Fiber
Kaylo 10	Owens Corning Fiberglas	Glass Fiber
Micro-Lok 650	Johns-Manville	Fiberglas
Pabco Super Caltemp Type NA	Fibreboard Co.	Diatomaceous Earth, Non-asbestos Fiber
PF-CG	Owens Corning Fiberglas	Glass Fiber
PMF	Jim Walter Resources, Inc	Glass Fiber
Ruberoid Fiber Glass Pipe Insulation	Ruberoid	Glass Fiber
Snap-On	CertainTeed	Fiber Glass
TGA-1000	Alpha Associates	Tedlar and Glass Fiber
Therma-K	Ehret Magnesia	Glass Fiber
Thermashield	Technit	Ceramic Fiber
Transifoam	Johns-Manville	Polystyrene
Transitop	Johns-Manville	Wood Fiber
Uni-Jac	Pittsburgh Corning	Glass Fabric
Vapo-Lok	MMM Div. Inssular Products Corp.	Expanded Polystyrene
VB-Vapor Barrier	Johns-Manville	Kraft Paper with Glass Fiber
VBC-Vapor Barrier	Johns-Manville	Glass Fiber
VC	Riva and Mariani	Cellular Glass
Z-Lock	Fibreglas, Ltd.	Glass Fiber

2. Substitute Materials for Sprayed-on Asbestos Insulation

TRADE NAME	MANUFACTURER	SUBSTITUTE MATERIALS
1. Cafco	USM	Mineral Fibers
2. Cafcote H	USM	Mineral Fibers
3. Cermafiber	Spraycraft corp	Ceramic Fiber
4. Ceamwool	Johns-Manville	Ceramic Fiber
5. Encacel	Childer Product Co	Ceramic Fiber
6. Encacel V	U.S Rubber Co	Urethane Polyvinyl
7. Ensolite	U.S. Rubber Co	chloride Polyvinyl
8. Ensolite Type	National Cellulose Corp	chloride Cellulose

3. Substitute Materials for Asbestos-Containing Panels or Wallboards

TRADE NAME	MANUFACTURE	SUBSTITUTE
1. Bestwall	Georgia Pacific	Gypsum Calcium
2. Cal-Shake	U.S. Gypsum	Expanded Perlite
3. Caretemp	Celotex	Polystyrene
4. Cellofoam	USM	Cellulose
5. Cellutron	Owens Celotex	Perlite Ceramic
6. Celot-Therm	USM	Fiber Ceramic
7. Ceramfab	Keene corp.	Fiber Ceramic
8. Delta-T	Dow	Fiber Molded
9. Doraspan	Sinclair-Koppers	Foam Wood Fiber
10. Dylite	Sinclair-Koppers	Glass Fiber
11. Econacoustic	Alpha Associates	Treated Cotton
12. Filomat-D	Cotton, Inc.	Gypsum
13. Fire Stop	Johns-Manville	Polstyrene Foam
14. Firetard Type	USM	Glass Foam
15. Foamgrid	Pittsburgh Corning	Polyurethand Form
16. Foamsil-28	Pittsburgh Corning	Polystyrene EPDM
17. Foamthane	Armstrong Cork co.	& Aramid Gypsum
18. SE Armalite	Dow	
19. Styrofoam	Johns-Manville	
20. Watertbacker		

4. Substitute Materials for Asbestos Containing Fabrics

TRADE NAME	MANUFACTURER	SUBSTITUTE MATERIALS
1. Alpha-Maritex 311-RW	Alpha Associates	Fiber Glass
2. Alpha-Maritex # 84205	Alpha Associates	Glass Fiber
3. Aramid	DuPont	Synthetic
4. Cell-tile	Ceel-Co.	ABS Plastic
5. Cerafelt	Johns-Manville USM	Cermamic Fiber
6. Ceramfab	USM	Cermamic Fiber
7. Fiberfraz	Carborundum Co.	Cermamic Fiber
8. Fiberseal	Pyrotek, Inc.	Glass Fiber
9. FireStop	Cotton, Inc.	Treated Cotton
10. Flexfelt	General Insulating	Rock wool
11. Flextra	Raybestos-Manhatton	Cotton/Aramid Blend
12. Fyrepel	Fyrepel	Glass Fiber
13.Glas Ply	Johns-Manville	Glass Fiber
14.Flassbestos	Raybestos-Manhattan	Glass Fiber
15.Glass Web	Steiner Industries	Glass Fiber
16.GVB Glass Cloth Vapor	Johns-Manville	Glass Fiber
17.Hansoquilt	Baldwin-Ehret-Hill	Glass Fiber
18.Insulfas	Benjamin foster	Glass
19.Kynol	American Kynol Corp	Novoloid Fiber
20.Nomex	DuPont	Synthetic
2 1 Nor-Fab	Hitco (Amco)	Synthetic
22.Nor-Fab	AMATEX Corp.	Synthetic
23. Preox	Gentex Corp.	Heat-Stabilized Polyacrylonitrile
24. Pyroglas	Raybestos-Mahattan	Glass Fiber
25.SF 2600	Santa Fe Textiles, Inc St.	Ceramic fiber
26.Sisalkraft	Regis Paper co	Kraft Paper/Aluminum
27.Snap Form	CertaiTeed	Polyvinyl Chloride Glass
28.Tempo	Tempo glove Mfg	Fiber or Leather Kevlar
29.Terrybest	A-Best Co	Perlite
30.Themafiber	U.S. Gypsum	Kevlar & Other fibers
31.Themobest	A-Best Co	Ceramic Glass
32.Thermo-Ceram	Garlock Inc	Fibers Glass
33.Themoglass	AMATEX Corp	Fiber
34.Themo-Sil	Garlock, Inc.	Polystyrene
35. Zonolite Dyfoam	W.R. Grace Co	

5. Substitute Materials for Asbestos Containing Materials

TRADE NAME	MANUFACTURER	SUBSTITUTE MAT
Alumino-Hi-Temp	Carey	Alumina
Careytemp 1500	Celotex	Expanded
Cem-Fil	Asahi Glass Co.Ltd	Perlite Glass
Cerablanket	Johns-Manville	Fiber
Cerachrome	Johns-Manville	Ceramic
Epitherm	Eagle-Picher	Fiber
1200	Nonco Corp	Ceramic
Feldina	Johns-Manville	Fiber
Fesco	Keene Corp.	Polyvinyl
Board	Celotex	Chloride
Mono-	Celotex	Non-Asbestos
Block	Armstrong Cork Co.	Mineral
MW-One Insulating Cement	Fibreboard Corp.	Perlite
MW-50		Mineral
Nonpariel	Fibreboard Corp.	Wool
Pabco Super Caltemp Type		

6. Substitute Materials for Brake Linings or discs.

TRADE NAME	MANUFACTURER	SUBSTITUTE
Aramid	DuPont	Synthetic
Kynol	American Kynol Corp.	Novoloid
Metal-Might	Lear Siegler, Inc.	Fiber
Premium	Euclid Industries	Metallic
Scan-Pac	Scan-Pac	Fiber
Star Line	Abex Corp.	Synthetic

Objective: To provide an overview of the contract specifications used for asbestos abatement projects.

Learning Tasks: Information in this section should enable participants to:

- Recognize the importance of well-designed, detailed contract specifications.
- Become familiar with the key elements of contract specifications.
- Recognize the basic components of material, equipment, and substitution specifications.
- Become familiar with the importance of detailing specification for the execution of work.
- Further recognize the need for interdisciplinary approaches to asbestos abatement.

CONTRACT SPECIFICATIONS

Well-designed, detailed contract specifications provide the overall guidance for each asbestos abatement project. These specifications permit the contractor to provide the building owner or designer with an accurate estimate or bid for completing the project. With few exceptions two contracts are required for each project. One contract is established with the contractor performing the actual abatement work, and a second contract between the building owner (or designer) and the air sampling professional.

Poorly designed specifications will result in a poorly performed project. If details are omitted in the specifications or procedures are unclear, the bids will vary tremendously. Likewise, Contractors must spend the necessary time to read the specifications in their entirety before the pre-bid walkthrough of the intended job site. The National Institute of Building Sciences (NIBS) "Guide Specifications for Asbestos Abatement Projects" may be a helpful reference for individuals designing projects. With the enactment of AHERA in 1987 and the issuance of the Asbestos School Hazard Abatement Reauthorization Act (ASHARA) in 1992, most asbestos response actions in public and private schools (K-12) as well as public and commercial buildings must be designed by an accredited asbestos abatement project designer. Many states have instituted minimum educational and experience requirements for project designers.

CONTRACT SPECIFICATIONS

This section is written primarily as a guide to preparing specifications for a large asbestos abatement project in anticipation of obtaining bids from a select group of qualified contractors. However, for small projects, many facilities obtain asbestos abatement services through a "work order" or "purchase order" system. This mechanism can work very well if the contractor is well qualified and if work practices to be used are clearly understood. It is highly recommended that the guidelines in this section be used to develop a basic specification by which all work order projects are done at a given facility. This specification can then be used as a basis for negotiating unit prices with work order contractors.

Contract specifications (specs) are a written set of standards and procedures informing the contractors or abatement professionals of materials and operations necessary to successfully complete an abatement project. Typically, these documents cover the entire spectrum of an abatement project; from site investigation by the contractor to use and application of replacement materials.

The specifications are usually prepared in book form, and with the contract drawings and any addenda or change orders constitute the contract documents.

During preparation of the contract documents, conflicts may result between specs and drawings. In a case such as this, specs will take precedence over drawings, unless otherwise stated in the contract documents.

During the bidding period, in which each contractor determines his/her cost of performing the work, quite often it is necessary for the designer to change or alter the project specs. Such a change is issued as an addendum to all bidding contractors in order to ensure that all parties are bidding on the same information. Basically, an addendum is a legally incorporated update to the drawings and/or specs prior to submittal of bids.

Should any change be necessary in drawings or specs after the contract has been awarded, a change order is issued. This legally binding action is signed by the owner, designer, and contractor. It may entail an increase or, rarely, a decrease in the dollar amount of the contract, and may increase or decrease the amount of time necessary to perform the work.

Information which is often included in contract specifications is contained on the following pages. It is important for an asbestos abatement professional to keep in mind that no two abatement projects will be identical. Various aspects of a project will be similar from job to job, but no one set of contract specifications can be used from project to project without modification or large-scale changes. Hence, the designer who will develop the specifications and most likely represent the interests of the building owner will want to become familiar with all aspects of the project.

ELEMENTS OF THE SPECIFICATIONS

Site Inspection

Contract specifications issued to bidding contractors will usually include general information on bidding requirements. Often, in order to become a qualified bidder, a site inspection will be required of the contractor. This is done to assure all parties that the contractor has become familiar with the conditions of the project, including the physical condition of the site, access to water and electricity, as well as the character and quantity of the ACM involved in the abatement project. Regardless of whether or not a site investigation is specified, a prudent contractor will want to become familiar with the project in order to present a responsible estimate for the work. Associated with the site inspection

is a pre-bid walkthrough and meeting during which details of the project will be discussed, and the competing contractors can ask questions to clarify any misunderstandings.

Scope of Work

The project "scope of work" will be detailed in the specifications. This section will include a description of ACM locations and quantities to the extent they have been determined (usually also provided on drawings), the type of abatement procedures to be used in a particular area, timeframe for project completion, and any restoration requirements that may be necessary.

Description of Work

A "description of work" section will detail abatement measures for each work area. Additionally the contractor will be required by this section to supply all labor, materials, services, insurance, equipment, etc. necessary to carry out the work in accordance with the specs and all applicable laws.

Any special conditions that may be encountered on the project (i.e., high temperature steam lines, operational equipment, etc.) will be detailed. This section also may include the requirement that the contractor restore an abatement area to conditions equal to or better than original. The contractor will be held responsible for any damages caused during the course of his work and will remedy any damages at his own expense.

Submittals and Notices

Submittals and notices are important in getting the abatement project off to a smooth start. The contract specs should spell out who is responsible for properly notifying applicable regulatory agencies, in addition to securing the necessary permits for waste handling and disposal procedures. Documentation that the contractor's supervisors, foremen, and workers are properly trained, licensed where required, and medically certified under applicable regulations will also be submitted to the building owner. It is also important that any existing damage be documented by the contractor and submitted to the owner prior to any work commencing. This will protect both the contractor and owner from punch-list disputes at the close of the project.

Included in the contractor's submittals will be a list of equipment to be used along with any certification documents which the specs call for. (For example: HEPA- vacuums conforming to American National Standards Institute (ANSI) standards). This will include respirators as well as other equipment for the project. During the abatement activity itself, the specs may call for weekly progress reports on abatement status, transport manifests, waste disposal receipts, project monitor logbooks, bulk and air sample results, and documentation of HEPA-filter changes.

In addition to requirements for the contractor, the project specifications will sometimes require the building owner to perform certain functions. This may include notification of building occupants of the work to be performed and making arrangements for temporary relocation. Additionally, the owner will usually have to make available to the contractor the results of any pre-abatement air sampling or bulk sampling tests.

Site Security

An area of concern for building owners and contractors alike is site security and emergency planning during the abatement project. Contract documents will specify that only authorized personnel will be allowed access to the work site (employees of the owner and subcontractor, air monitoring personnel, inspectors). This is done to limit the liability of the parties involved and to prevent any worksite problems. Emergency planning will include written notification to police, fire, and emergency medical personnel regarding the proposed abatement activity, and the proper procedures to follow if an emergency situation should arise.

MATERIAL, EQUIPMENT, AND SUBSTITUTION SPECIFICATIONS

Material Specifications

Material specifications will include documentation of onsite storage regulations, details of material to be used (example: black polyethylene for worker decontamination units), any special materials to be used to protect objects in the work area, and proper containers for asbestos waste. Materials for removal (surfactants), encapsulation, and any replacement materials will be specified here. Proprietary specifications will require the use of a specific material from a specific manufacturer, whereas non-proprietary specifications will indicate materials or performance requirements and will allow the selection of materials which will perform to those limits.

Equipment Specifications

Equipment specifications will detail the performance requirements of units such as negative air filtration units, Type-C respirators and associated compressed air systems, and protective clothing and safety equipment such as footwear, hardhats, and eye protection. Equipment used in the actual removal of ACM will also be specified, such as scaffolds, ladders, sprayers and their capacity ratings, and any other equipment deemed essential for proper completion of the job.

In the event that a substitution of materials or equipment is necessary, specifications will usually call for the submission of complete technical data and information on the substitution from the contractor to the building owner or his/her representative. The building owner will then review the information and have the authority to approve or disapprove such substitution. It is important to keep in mind that the terms "or equal" and "or equal as approved by owner" do not mean automatic owner approval of substitutions. Any material or equipment that the contractor feels is equal to what has been specified should be approved by the owner before being used. There have been many instances where a contractor has had to redo work when unapproved substitutions were found to have been made.

SPECIFICATIONS FOR THE EXECUTION OF WORK

Procedures for preparation of the work area will be specified in the contract documents. The specs will include the shutting down and locking out of electrical power to the area, if possible. Provisions for temporary power and lighting may have to be made, and are usually the responsibility of the contractor. The HVAC system of the building will also need to be shut down or modified to prevent the spread of contamination to other areas of the building. This is an especially important consideration in buildings where adjacent areas will remain occupied. Considerations will be detailed for furniture, machinery, etc., which will be cleaned, removed from the work area, and stored. Any items not able to be removed will be cleaned and sealed using specified methods.

Specifications will often detail the construction design of the decontamination facilities, including location, size of chambers, and number of airlocks. Additionally, entry and exit procedures will be set forth, with provisions for waste material removal and storage.

The work area will be prepared according to project specs which will usually state polyethylene sheeting thicknesses, layers, and seam overlap. Special considerations for equipment which cannot be shut down and removed from the work area (computers, communications switching equipment), and for work areas with adjacent occupied space will also be specified. It is vital that contractors as well as other abatement professionals become familiar with any special circumstances.

Once the preparation of the work area is complete, specifications will usually detail the methods and order of removal of contaminated materials in the work area. Any encapsulation or enclosure of asbestos-containing material will be specified. As removal, encapsulation, and/or enclosure operations are completed, the waste material will be taken out of the work area, transported by a licensed waste hauler, and disposed of in an approved landfill. Proper documentation and retaining of receipts from disposal practices are specified to be delivered to the building owner.

RESPIRATORY PROTECTION

Specifications for respiratory protection may vary with either the type of abatement activity or with fiber concentrations. HEPA-filter cartridge respirators may be specified for use during work area preparation and/or following complete removal of gross contamination and one or more phases of cleaning. Type-C air supplied respiratory protection is usually specified from the beginning of actual ACM removal until gross contamination has been cleaned-up and removed from the area. Powered air-purifying respirators are sometimes specified in place of cartridge respirators.

When considering airborne fiber concentrations and specifications for respirators, a general rule of thumb is to provide a protection factor that would maintain a fiber concentration of 0.01 fibers per cubic centimeter (f/cc) inside the mask. Due to daily fluctuations in fiber concentrations and analytical considerations, it is often difficult to ensure that a concentration of 0.01 f/cc is maintained. Increasingly, a combination of fiber concentrations in the work area and abatement activity is used in specifications to determine the level of respiratory protection. For example, during the second phase of cleaning, PAPRs may be acceptable if measured fiber concentrations would justify that adequate protection was achieved. Documentation that workers have received proper training in the purpose and use of respirators and that workers have been qualitatively or quantitatively fit-tested should always be specified.

CLEARANCE AIR TESTING

Specifications should include requirements that a complete and thorough visual examination of the work area be conducted by the owner's representative prior to final air clearance testing. The procedures to be followed for final air clearance testing should always be thoroughly spelled out in the specs. Sampling locations and the number of samples to be collected will sometimes be specified. Often, however, this is left to the discretion of the air sampling professional. Sampling methods (aggressive or non-aggressive), analytical techniques (PCM or TEM), and sampling times will be specified according to applicable regulations. The specs should also state very clearly what will be considered "clean" in regard to airborne fiber levels.

WASTE DISPOSAL

Contract specifications will indicate an EPA authorized landfill that will accept asbestos waste as an acceptable dump site for the abatement project's waste. A recordkeeping format whereby all waste shipment records required by the NESHAP regulations, receipts, and manifests will be retained and delivered to the building owner will usually be included.

Specific requirements for the transportation of asbestos waste will include preparation of the waste containers and truck, proper loading and unloading procedures, personal protective equipment that must be worn by waste handlers, and decontamination procedures for the transport vehicle itself.

MATERIAL REPLACEMENT

Replacement of materials which have been removed, such as fireproofing, insulation, a drop or suspended ceiling, or any renovation work that incorporates new material will be completed following passage of clearance testing. This work may be covered under the original abatement contract and included in the same set of specs, or may be under separate contract and specifications. Following the completion of replacement, or if no additional work is to be done, the contractor will be required to remove all critical barriers, re-secure any objects or fixtures taken from the area, replace furniture and fixtures in their former positions, and re-establish the HVAC, electrical, and any plumbing systems to their proper working order. Finally, any damage resulting from the contractor's abatement activities will be repaired.

OTHER ABATEMENT PROFESSIONAL

Contract documents will specify the qualifications and responsibilities of other abatement professionals associated with the particular project. These other persons will include the Asbestos Project Manager – usually the owner or owner's representative. The Project Manager will be responsible for assisting in decision making, developing, implementing, and enforcing the contract specs, inspecting the work areas and critical barriers, and possibly coordinating bulk and air sample collection as well as other duties.

The Air Sampling Professional will be in charge of conducting air sampling in accordance with the project specifications. The types of sampling–pre-abatement, area sampling, personal, ambient – and any air volume requirements or associated sampling strategies will be outlined. The procedures to be followed for final air clearance testing will be detailed in accordance with applicable Federal and State regulations.

The laboratory services utilized will be specified as having to be accredited for both bulk and air sample analysis. Turnaround time for samples may also be specified, as well as accreditation requirements for individual analysts.

A well-designed and organized set of contract specifications will provide for a successfully completed abatement project. However, it is important for a contractor to remember that not all regulations and requirements will be included in project specs. For this reason, a contractor should become familiar with any Federal, State, and Local laws that apply to his or her situation.

Objective: To provide a brief awareness of the legal and insurance issues affecting asbestos abatement design and project completion.

Learning Tasks: Information in this section should enable participants to:

- Grasp an overview of common and statutory law, including tort litigation and regulatory compliance.
- Discover the value of recordkeeping and documentation.
- Appreciate asbestos abatement contract specifications and documents.
- Understand bonding, workers' compensation, and insurance requirements.
- Avoid legal pitfalls on asbestos abatement projects.

LEGAL PROBLEMS IN ASBESTOS ABATEMENT

The owner or manager of a building facing the presence of asbestos has many potential problems. A major area of concern for building owners is the legal implication of asbestos present in buildings. In general, the problems can be divided into two areas. One broad area is the law of torts or legal wrongs. This could include the possibility of an owner being sued for failure to properly abate or contain asbestos. It could also include a cost recovery action by an owner who has or will perform an abatement program. The second broad area of concern to owners is the area of contracts as related to asbestos abatement programs. This is the area of emphasis in this section.

TYPES OF CONTRACTS

AIA THE AMERICAN INSTITUTE OF ARCHITECTS

Once a building owner has decided to begin an asbestos abatement program, the owner is faced with many practical questions. The first of these relates to contract documents. A typical contract used for this type of work is the American Institute of Architects (A.I.A.) form contract. This will frequently consist of A.I.A. Form A101 ("Owner-Contractor Agreement Form – Stipulated Sum"), a four-page document which typically contains the name of contracting parties, the contract amount, the start and completion date, and other general data. Accompanying the A.I.A. Form A101 is the A.I.A. Form A201. This document which contains much of the "boiler plate" language. A.I.A. A201 has been around for many years. Its clauses have been frequently litigated, and lawyers and building owners are familiar with the normal interpretation placed on the language in A.I.A. A201. This provides for a degree of certainty which is desirable. Of course, disputes do arise under these clauses. However, it is much easier for counsel, a building owner, and a contractor to resolve differences in this known area than in the unknown areas of a job specification on an asbestos abatement contract.

The third document that will typically be included in an asbestos abatement contract is the job manual or job specifications. Whether the job is private and an A.I.A. contract is used, or whether the job is public and an A.I.A. contract is used, or whether the job is public and, for example, the General and Special Conditions, a two-inch thick book of standard specifications on a Federal project, is used, any asbestos abatement project requires a set of job specifications. In addition, the National Institute of Building Sciences (NIBS) has produced a guide specification, which may be helpful too. The fact that these job specifications have not been through the courts and have not been subject to the interpretations placed on the standard specifications in A.I.A. forms and other standard forms like the Federal forms, makes architects and engineers cautious in their interpretations of asbestos abatement specifications – and it should.

Frequently, architects and engineers give very strict interpretations to these clauses and are much less permissive than in interpretation of other clauses involving less hazardous activities.

CONTRACT SPECIFICATIONS

One of the primary important areas in the specifications is the work description. Typically, if a whole building is involved, the work description may be no problem. However, if a contractor is being asked to remove only a segment of a floor of a building, or the like, a serious abatement problem may exist. When the plastic barrier is removed, will the "clean air" obtained still be clean? Another problem is illustrated by the job where the contractor was asked to remove friable material from beams over a dirt basement in a commercial building. Specifications did not call for barrier protection for the dirt. The contractor, fully in compliance with the specifications, removed the asbestos and contaminated the dirt. The result was that a second abatement effort was required for the removal of several inches of dirt in the basement. This cost, quite unnecessary, was the result of inadequately prepared specifications. Of course, drawings are of great assistance in properly defining job boundaries and the specifications.

Another issue that frequently arises is that of furniture, fixtures, and equipment. Most asbestos abatement projects require the removal and then reinstallation of furniture, fixtures, and equipment. The owner should make a realistic assessment of whether salvaging of fixtures is feasible. The age and condition of the equipment must be assessed. It will only promote contractual disputes for the owner to insist that a contractor remove and reinstall fixtures which will be damaged or destroyed by that very process. The owner should not expect to engage in a "backdoor" renovation of his building at the contractor's expense, by placing specification requirements on the contractor which are unattainable. From a contractor's point of view, a pre-abatement job inspection is vital to determine whether or not the specifications are realistic.

SITE SECURITY

Site security is another issue that should be specifically addressed. In certain settings, such as a school or hospital, an insecure job site can have grave legal implications. Of course, any construction site can be dangerous but an unmonitored asbestos abatement site may be an "attractive nuisance" for children or others. The issue of whether security must be maintained on a 24- hour basis is also one which must be addressed. Again, this is an area where more attention must be given if only one area of a building is being abated while other areas are occupied. If an entire building can be isolated, security is a much easier process. Certainly the regulatory warning signs required by the U.S. Occupational Safety and Health Administration (OSHA) are important; however, these should be viewed as a minimum warning and further warnings of security are frequently necessary.

EQUIPMENT

Each owner must participate, if only through his architect or engineer representative, in equipment requirement decisions on an asbestos abatement project. The decision about the number of negative air pressure units, what type of respiratory protection equipment is required, what type of clothing is sufficient, what number of high efficiency particulate air (HEPA) vacuum units are necessary, and other similar equipment requirements are very important to the success and safety of the job. OSHA requirements again should be viewed as minimal requirements. If one is not going to employ the best available technology, that decision must be justified. There may be instances when Type "C" respiratory protection is not required, such as during glovebag projects. The counterbalancing danger of employee accidents from the hoses on scaffolding might possibly outweigh the necessity for an air-supplied respirator. The owner and his representative must be prepared to justify such decisions. Of course, if the decision to employ less than maximum protection is justified, cost is a reasonable factor to be considered.

INSURANCE

A major issue for asbestos abatement contractors is insurance. At present, the insurance "crisis" is easing and contractors who have not been able to obtain liability insurance now have it more readily available. From the owner's or contractor's perspective, it is first important to understand what the insurance requirements may be. The first issue is who must be protected. Generally speaking, the owner will want to require protection for himself, the contractor and the workers of the contractor. This will mean that the owner will wish to be an additional insured under the contractor's policy and to be furnished a certificate of insurance. It is important that the certificate of insurance specify the types of coverage afforded. If the certificate of insurance is not completely clear, the owner should request a copy of the contractor's insurance policy. Generally, it will be better practice to review the contractor's policy. This is particularly true with asbestos work exclusions now appearing in such policies. In lieu of a review of the policy, an affirmative representation from the insurer that the work includes asbestos abatement coverage may suffice; however, in most cases, a legal opinion will be necessary.

Insurance must protect the insured from both personal injury and property damage claims. While this is standard in liability policies, in personal injury claims, there may be an exclusion of coverage for asbestos-related claims and in property damage coverage, there may be exclusion for the "pollution hazard" which could encompass asbestos fibers released onto the property of another. Therefore, the owner must be sure that he is protected from the real hazards presented by asbestos abatement. Of course, it is necessary to insure against the normal construction hazards as well.

Another very important point is to determine for what length of time the insurance will be in force. Clearly, the insurance should be in force during the contract. However, it is most important that the coverage continue after the contract is completed. This will ensure that, if the contractor fails to completely abate the asbestos problem and the owner fails to discover the contractor's oversight, the owner will nonetheless be protected.

Closely related to this is the need to determine whether the insurance is "claims made" or "occurrence" coverage. "Claims made" coverage will insure one for claims made during the policy period. "Occurrence" coverage will insure for an occurrence taking place during the policy period, whether or not the claim is made during that time. This is very important because of the long latency period associated with asbestos diseases. If the insurance is "claims made" insurance and the policy expires or lapses, the owner may not be protected many years later when the claim is actually made. This matter should be carefully reviewed with the insurance representative and with legal counsel.

Generally speaking, the owner needs liability insurance coverage for both personal injury and property damage risks, a builder's risk policy which protects against the particular risks occurring during construction and of course, workers' compensation coverage which complies with the law of the state where the work is to be done.

In most states, workers' compensation covers injuries received on the job by employees. Firms engaged in asbestos abatement activities should be aware that some states do not cover occupational disease that arises many years after employment has terminated. Further, some states permit lawsuits to be filed against the employer by its employees. Since this complex issue exceeds the scope of this manual, employers involved in asbestos abatement should include workers' compensation when consulting with their legal counsel.

One other important aspect of specification writing and interpretation is the clean air standard to be used. The safest legal path is going to lie with the best available technology. According to many industrial hygienists and other professionals, this requires the use of Transmission Electron Microscopy (TEM) to clear asbestos abatement projects. If one is going to conduct an asbestos abatement program most specifications will call for the use of TEM.

Asbestos Supervisor Skills

Asbestos Supervisors are, by definition, **leaders in the asbestos workplace.** Their duties are to **supervise all types of asbestos abatement works**; effectively **managing site paperwork** with **competent site management skills** which include health & safety issues; and exhibit **confident communication skills to effectively manage** a team of asbestos operatives.

A supervisor **must be able** to : 1) **plan and execute jobs**; 2) produce take-offs and **order materials** based on calculations of project specifications; 3) **monitor tools/equipment**--coordinate with warehouse to procure, secure, mobilize, repair and replace as needed; 4) **supervise and mentor all work** performed on assigned jobs; 5) **direct all technicians' work activities** to meet budget, performance and safety requirements: 6) **work with crew** with a hands-on approach-- assist as needed; 7) **communicate clearly to all assigned to the job all requirements**-- including the schedule, the execution plan, abatement/remediation techniques, equipment usage, and safety/emergency procedures:8) **train/mentor employees** on job techniques, safety procedures and equipment use; 9) **report regularly to management** on daily operations of the project; 10) **communicate** with customers agents, other contractors, subcontractors and others: 11) **prepare asbestos reports** using Microsoft Office.

A supervisor **may be required** to: 1) **sample suspect asbestos-containing materials** in accordance with AHERA guidelines: 2) **sample suspect lead-based paint or** mold in accordance with applicable regulations or protocols; 3) **perform oversight and air monitoring** during asbestos, lead paint or mold abatement projects; 4) **calibrate field-monitoring equipment**; 5) assist in **data tabulation and preparation of reports**; 6) **assume safety responsibilities** including: a) completing **safety orientation** program; b) abiding by **safety and personal protective equipment policies**; c) demonstrating **safe work practices** and avoiding unsafe behaviors; d) **driving defensively**; d) **avoiding preventable accidents**; e) always **wearing a seat belt**; f) **attending regularly scheduled safety training**; g) **reporting all unsafe conditions**; and performing **other duties** as assigned.

SUPERVISION AND TRAINING

The heart of an asbestos abatement project is not the equipment, although the equipment is important, or the physical structures, although they are vital, too. The heart of the project is the care and skill exercised by the workers who remove the asbestos-containing materials. If this occurs in a professional manner, the owner can expect a good result.
If the work is sloppy, good equipment will not save the job.

Some projects are conducted with both trained and untrained workers. Many asbestos abatement contractors hire local workers and train them. Unfortunately, a few contractors hire local workers and do not train them. From the owner's perspective, it is vital that the job superintendent be an experienced asbestos abatement worker. He or she must have had the experience on various projects and under various conditions. The owner should require documentation of the experience. It will behoove the owner to check on the quality of the preceding jobs. The superintendent is the key to the work.

Many jobs are sufficiently extensive to require a job superintendent and a job foreman. The latter person typically will be in the barriered area actually supervising the workers while the job superintendent may be in and out of the area at various times. If a job foreman is required, he or she too, should be experienced and able to instruct the workers on-site and personally supervise actual preparation, removal, and follow-on clean-up activities. It is desirable to contractually require that both of these positions be filled by qualified individuals.

Many owners are now asking for contractors with experienced work crews. It is possible to contractually require such, although the price of the job will probably go up. From a legal standpoint, these workers, particularly if untrained, are the weakest link in the chain. If the workers hired by the contractor are untrained, it is essential not only to meet such OSHA requirements as medical examinations, but also to conduct a program of worker training.

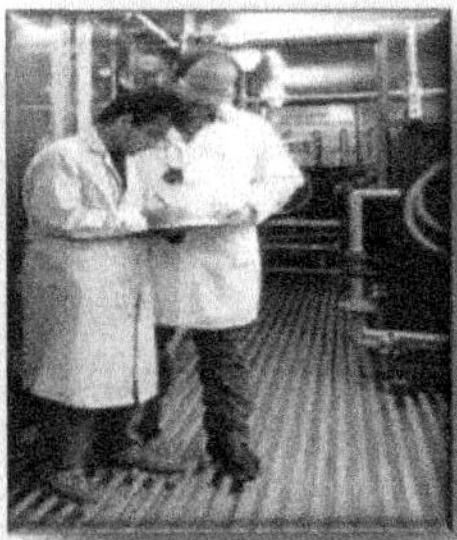

The U.S. EPA Asbestos Hazard Emergency Response Act (AHERA) Regulations stipulate that all abatement workers be accredited by completing at least a 4-day EPA approved training course, including lectures, demonstrations, hands-on training, respirator fit-testing, and a course examination. OSHA also has requirements for minimum training based on the type of asbestos activity a person will perform. From a legal standpoint, it is most important that records be kept to document training was given to each worker.

Many contractors require testing of their workers before they allow them to proceed to the job site. If a test is administered, the results should be retained. Records of all training should be kept for each worker. The owner should be permitted to inspect these records and copy them if desired. In fact, the owner may wish to maintain a complete file on all workers who worked on his job. This could avoid the problem of the contractor going out of business later and an asbestos claim arising against the owner many years thereafter based upon an alleged failure to warn, or the like. (See the "Regulatory Review" section for more information on the required training for both EPA and OSHA.)

Not only are federal requirements for training and accreditation important, but many states have licensing requirements of their own. It is important that individuals also comply with all state and local regulations pertaining to licensing when conducting abatement work.

RECORDKEEPING

This area points to the need to maintain permanent records on all phases of the job. It is not enough to maintain these records for a few years. OSHA has specific requirements for how long certain records must be maintained. Asbestos disease latency periods may extend 30 or more years beyond the work date. These records should be stored and maintained permanently.

ADEQUATE TIME FOR JOB PERFORMANCE

One of the most difficult problems in asbestos abatement work is that of time. Once the owner has decided to conduct an asbestos abatement program, he is virtually always in a hurry. Many programs are specified with very short time limits of two to five days.

To avoid contractual disputes, it is desirable for the owner to specify enough time to allow the project to be completed. Questions of access, other contractors, and the owner's employees must be considered. Generally speaking, qualified abatement contractors will proceed quickly once they begin a job. Owners are frequently critical of late starts resulting in late finishes. It may therefore be in the owner's interest to provide some penalty for a late start as a means of emphasizing to the contractor that the start date is important.

On the other hand, owners are frequently guilty of not providing the site at the time specified. This may result in difficulties for the contractor with his other work. The resolution is to cooperatively come up with a time for the project which will be realistic both in length and in calendar placement for the owner and the contractor.

As to delay damages, many owners choose to use liquidated damages as a vehicle to make sure that contractors finish the job. If extensive renovation work or other important use of the facility is contemplated, the liquidated damages may not suffice to cover the actual damages. In these instances, the owner may find that he is limited to damages that are less than what he wants. Most courts have limited the owner to liquidated damages if they are specified on the basis that the reason for specifying liquidated damages is the uncertainty of calculation of actual damages.

DISPOSAL OF ASBESTOS-CONTAINING WASTE

Another important issue is that of materials handling. The contract should provide information indicating the asbestos material must be properly containerized, labeled, and shipped to an approved dump site. The owner should obtain receipts to show the delivery of the materials. It will be desirable to have an inventory of the drums to compare to the dump site receipt. This kind of attention to detail will eliminate possible claims of improper toxic substance dumping.

USE OF IN-HOUSE PERSONNEL

Many owners are public entities and have severe budget constraints. There is a temptation, therefore, to use in-house staff for important functions in an asbestos abatement program. Use of in-house staff in lieu of outside experts may be a serious mistake and may result in legal liability if the in-house staff is unqualified. AHERA regulations require that all maintenance and custodial personnel receive at least 2 hours of awareness training, with an additional 14 hours of training mandated for personnel whose work activity may disturb asbestos. Issues of sovereign immunity and waiver of sovereign immunity are always, of course, involved. An outside expert is probably going to be desirable and needed if the public agency has a serious enough problem to spend the time to thoroughly train an in-house staff member on operations and maintenance and the like. The expert will be hired on a contractual basis and will be able to train the in-house staff person.

SELECTION OF QUALIFIED CONTRACTORS

Another problem peculiar to public works is the low bid problem. Low bid may equate to an unqualified asbestos abatement contractor. The way to resolve this problem and still comply with low bid obligations of a public agency is to pre- qualify the contractor. Pre-qualification should consist of various requirements. These may include experience, training, formal education (for instance, attending seminars), insurance, as has been discussed, and a job inspection. A public agency simply cannot afford to hire an unqualified contractor on the basis that the contractor submitted the low bid. Also, if the contractor does not have the time to visit the site and inspect it, he probably doesn't have the time to do the job correctly.

Frequently, an asbestos abatement contract may be coupled with other renovation work. It is virtually always necessary to have some replacement structures for the asbestos-containing structures that are removed. Generally, it is going to be more desirable to have the asbestos abatement contractor function as a subcontractor under a general contractor in charge of the entire project. This will allow coordination of the trades. In a particularly dangerous job, it may be contractually necessary to require some training of other trade workers regarding the dangers of the asbestos work. Certainly, conferences between trade superintendents throughout the job are desirable and strict security is most important in a job of that type.

Using the prime contractor with an asbestos abatement subcontractor does raise a problem if the asbestos abatement subcontractor is unable to complete the work or is thrown off the job for some reason. In most instances, the prime contractor will not be able to complete the work himself. The owner may wish to reserve the right to select a substitute asbestos abatement subcontractor using the same criteria used in the original selection.

CLEANLINESS OF THE JOB SITE

One final note is that of the completion of the job by obtaining "clean air." Actually, the contractual requirements should be a combined requirement of clean air and clean surfaces. It is possible to have acceptable air monitoring tests and still have dangerous levels of asbestos-containing materials present within the building. In fact, once an abatement project is undertaken, the disturbing of the surfaces frequently makes the remaining material, if not removed, more friable. Therefore, the owner or his representative must be sure that the contractor has thoroughly removed all material <u>and</u> that the air is clean.

AIR MONITORING PROFESSIONAL

The integrity of the air monitoring person is vital. He or she should be qualified and experienced. There have been a few instances of falsified tests reported. An important contractual idea is to get air monitoring under a separate contract from the abatement contractor. An owner does not normally want to have a "turn key" contract where his abatement contractor supplies the air monitoring. The owner may wish the air monitoring paid for as part of the abatement contract, but should contractually provide for his own air monitoring with a separate firm or person.

BONDING

The difficulties in obtaining insurance have spread to the bonding industry. Traditionally, two types of bonds have been required in the construction industry to protect the owner or lender against the contractor's financial default:

- *payment bonds*, under which a surety company agrees to pay for labor and materials supplied to a project in the event the contractor fails to do so; and

- *performance bonds*, under which a surety agrees to complete performance of a project, if the contractor fails to do so.

Abatement contractors who have had their insurance canceled or not renewed experience difficulties in obtaining bonding. Bonding companies rely on the financial ability of the principal (the contractor) to respond to claims under payment and performance bonds. If a company is not insured against catastrophic liability, the financial underpinnings of the company are weakened, and the bonding company becomes apprehensive over issuing bonds. In a similar vein, lenders react adversely to the no insurance/no bonding problems of such companies.

Lenders advise companies who find themselves in such positions that lines of credit will not be renewed, for the same reasons given by the bonding companies. The difficulty being encountered by asbestos abatement contractors in obtaining bonding is severe. For reasons similar to those that caused the asbestos abatement insurance crisis, many contractors are unable to obtain sufficient bonding and, in some cases, any bonding. In addition to the general underwriting concerns about the contractor's ability to perform the work, another reason some bonding companies are unwilling to write bonds for asbestos abatement work relates directly to liability insurance problems.

Because the bonding contract often has requirements for the contractor to obtain and maintain certain liability insurance coverage on the project, the bonding companies fear that if the contractor has insurance problems, such as improper coverages or cancellation during the policy period, the potential loss that may otherwise be covered by liability insurance might be covered by the contractor's performance bond.

While the traditional concepts of bond underwriting may not be applicable to abatement contractors, it is nevertheless useful to understand them. The primary considerations of the bonding company in determining whether to bond a contractor are the ability of the contractor to perform the work and the contractor's financial ability. A proven track record of successfully completed projects, without ensuing litigation, is very helpful to the contractor in demonstrating to the bonding company its ability to perform the work. Financial stability is important not only with respect to the contractor's ability to perform the work, but also its ability to satisfy its indemnity obligation to the bonding company in the event a loss is suffered under the bonds.

Unlike insurance, a payment or performance bond gives the bonding company the right to recover back against the contract for any losses sustained by it under the bond. A somewhat more intangible, yet important, factor is the contractor's good character. Despite satisfactorily proving all of these items, however, a contractor may still not be able to obtain sufficient bonding in today's market. In such events, an owner may waive or refuse bonding requirements or arrange other contractual mechanisms to assure payment or performance.

There are numerous legal considerations involved in the evaluation of insurance and bonding coverage. The cost of insurance for asbestos abatement is significant, and if such expense is going to be undertaken, the coverage obtained should be satisfactory. While there are no easy solutions in this decision-making process, it is mandatory that contractors, consultants and owners undertake to become knowledgeable purchasers of insurance.

The shift in the types of coverages written for the contracting industry for occurrence to claims made, and the difficulty in obtaining bonds have placed greater emphasis on the contractor's commitment to the performance of work in a quality manner, the carrier's commitment to continuing to insure asbestos abatement contractors, and the quality of the carrier's coverage and insurance program in general. This makes the process of purchasing insurance more complicated, but a thorough review of the considerations outlined above will greatly assist the contractor, consultant, or owner in making a knowledgeable choice.

Objectives:

To provide an overview of the major asbestos- related regulations affecting abatement personnel.

Learning Tasks:

Information in this section should enable participants to:

- Become familiar with the general provisions of major federal asbestos-related regulations.

  DOT
  EPA
  OSHA
  MSHA

- Learn details of the major regulations pertaining to abatement projects and/or personnel.

INTRODUCTION

To date, two federal agencies have been principally responsible for generating regulations for asbestos control. These two agencies are the U.S. Occupational Safety and Health Administration (OSHA) and the U.S. Environmental Protection Agency (EPA).

Other federal agencies promulgating regulations on asbestos include the Department of Transportation (DOT) – regulations regarding the transport of asbestos; the National Institute of Standards and Technology (NIST) – establishing standards and protocols for laboratory accreditation, and the Consumer Product Safety Commission – banning asbestos in some products. These initiatives span the period of the early 1970s through the present.

A summary of OSHA and EPA regulations follows. Specifically covered are the OSHA Asbestos Standards; the EPA Worker Protection Rule; the National Emission Standards for Hazardous Air Pollutants (NESHAP); the Asbestos Hazard Emergency Response Act AHERA), the Asbestos Ban and Phase-Down Rule, and the Asbestos School Hazard Abatement Reauthorization Act (ASHARA).

OSHA®

U.S. OCCUPATIONAL SAFETY AND HEALTH ADMINISTRATION ASBESTOS STANDARDS

The Occupational Saf ety and Health Administration have established three sets of regulations that address asbestos exposure:

29 CFR 1910.1001	-	General Industry
29 CFR 1926.1101	-	Construction Industry
29 CFR 1910.134	-	Use of Respirators (General)

The construction-industry standard covers employees engaged in demolition and construction, and the following related activities likely to involve asbestos exposure:

- Removal
- Encapsulation
- Alteration
- Repair
- Maintenance
- Insulation
- Spill/emergency clean-up
- Transportation
- Disposal
- Storage of ACM

The general-industry standard covers all other operations where exposure to asbestos is possible, including exposure to occupants of buildings that contain ACM. In most cases, however, levels of airborne asbestos are not expected to reach the exposure standards in these buildings.

In general, OSHA coverage extends to all private-sector employers and employees in the 50 states and all territories under federal jurisdiction. Those not covered under the standard include self-employed persons, certain state and local government employees, and federal employees covered under other federal statutes. Persons engaged in inspection, management planning, and other asbestos-related work fall under OSHA's construction industry standard.

To enforce its standards, OSHA is authorized to conduct workplace inspections. In addition, employees have the right to file an OSHA complaint without fear of punishment from the employer. In turn, employees have the responsibility to follow all safety and health rules. OSHA may not conduct a warrantless inspection without the employer's consent. Citations are issued by OSHA during an inspection if the compliance office finds a standard being violated. The citation informs the employer and employees of the regulations or standards alleged to have been violated and of the proposed length of time for correction. Monetary penalties may also be imposed.

OSHA CONSTRUCTION INDUSTRY STANDARD

OSHA

The following highlights of 29 CFR 1926.1101 are condensed for easy reference. Participants are encouraged to become familiar with the OSHA standard as it appears in the *Federal Register.*

Exposure Levels

- Permissible Exposure Limit (PEL) – 0.1 fibers per cubic centimeter (f/cc), time weighted average (TWA). TWA means exposure concentration averaged over an 8-hour period.
- Excursion Limit (EL) – 1.0 f/cc as averaged over a sampling period of 30 minutes.

Asbestos-containing material is any material containing more than1% asbestos.

Asbestos work classifications:

Class I: Activities involving the removal of TSI and/or surfacing ACM/PACM

Class II: Activities involving the removal of ACM, which is not TSI or surfacing material

Class III: Repair and maintenance operations where ACM, including TSI and surfacing materials, is likely to be disturbed

Class IV: Maintenance and custodial activities during which employees contact but do not disturb ACM/PACM, and activities to clean up dust, waste and debris resulting from Class I, II, or III activities.

Competent Person is one who is capable of identifying existing asbestos hazards in the workplace, capable of selecting the appropriate control strategy, and having the authority to take prompt corrective measures. Personnel must be trained to meet the criteria of EPA Model Accrediation Plan for project designers or contractors/supervisors for Class I and II work, and training covering the topics listed in EPA's 16-hour O&M training program for Class III and IV work.

Negative exposure assessment is a demonstration by an employer that an employee's exposure during an operation is expected to be consistently below the PEL/EL.

Presumed asbestos-containing material (PACM) is TSI and surfacing material found in buildings constructed no later than 1980.

Regulated area is an area established by an employer to demarcate where Class I, II, and III asbestos work is being conducted, and any adjoining area where debris and waste accumulate. It also includes any area where airborne asbestos levels are anticipated to exceed the PEL.

Exposure Assessment and Monitoring

Initial Monitoring

Employers who have a workplace or work operation covered by this standard must perform initial monitoring to determine the airborne concentrations of asbestos to which employees may be exposed. This assessment must be conducted by a competent person before or at the initiation of the work or activity.

If employers can demonstrate that employee exposures are below the the excursion limit by means of

- Objective data, or
- Personal air sampling results collected from the previous 12 months, or
- Initial monitoring for the current job,

This is deemed a negative exposure assessment.

> **Note:** OSHA does not allow the use of objective data for exemption of monitoring of Class I work.

Periodic Monitoring

Periodic monitoring is required to be conducted daily within a regulated area for all Class I and II work, unless:

- A negative exposure assessment has been made, or
- All employees in a regulated area are wearing supplied air respirators operated in the pressure demand mode, or other positive pressure mode respirator.

However, if Class I work is performed using a control method not listed by OSHA in the regulations, or if a listed method is modified, periodic monitoring must be conducted regardless of the type of respirator worn.

If daily monitoring indicates that employee exposures are below the PEL/EL, then no further monitoring is required. Employees must be given the chance to observe the monitoring, and all monitored employees must be notified of the monitoring results as soon as possible.

Regulated Areas

The employer must establish a regulated area in those areas where 1) Class I, II, or III activities will occur, or 2) where airborne asbestos concentrations exceed, or there is a reasonable possibility that airborne concentrations of asbestos will exceed the PEL and/or EL. Only authorized personnel may enter regulated areas.

The following requirements apply to a regulated area:

- mark the area to minimize the number of persons within the regulated area and to protect persons outside the area;
- limit access to authorized personnel only;
- use respirators in accordance with paragraph H of the OSHA

Construction Standard;

- prohibit eating, drinking, smoking, chewing, and the application of cosmetics in the regulated area;
- competent persons must supervise work within the regulated area.

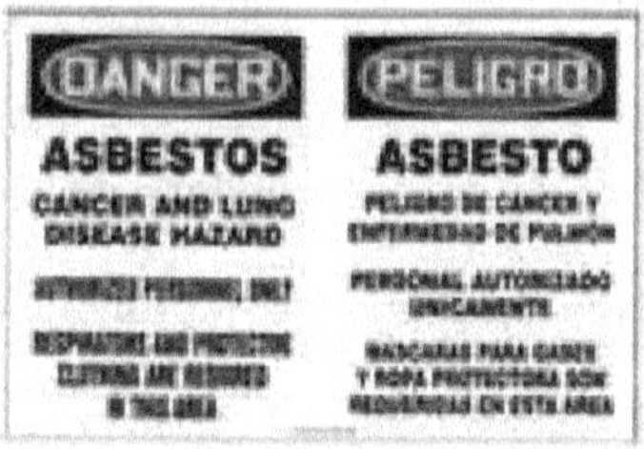

Warning signs must be displayed and must be posted at all approaches to regulated areas. These signs must bear the following information: In addition, where respirators and protective clothing are required within the regulated area, the warning signs must include the words:

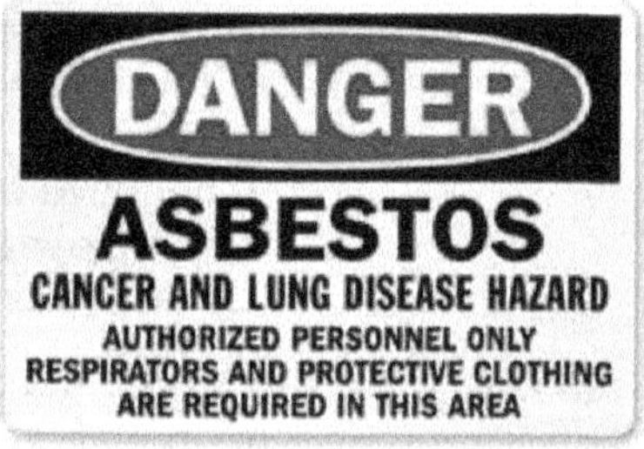

"Respirators and Protective Clothing are required in this Area."

The employer shall ensure that employees working in and contiguous to regulated areas comprehend these warning signs. This may be accomplished through the use of foreign language wording, pitographs, and graphics.

Warning labels must be affixed on all asbestos products and to all containers containing asbestos products, including waste containers. The label must include the following information and must be black, white, and red in color:

"Danger Contains Asbestos Fibers Avoid Creating Dust Cancer and Lung Disease Hazard."

Multi-Employer Worksite

When an asbestos project is being conducted at a multi-employer worksite, the employer performing work requiring the establishment of a regulated area must inform other employers of:

- the nature of the work,
- the existence/requirements of regulated areas,
- measures to protect the other employers' personnel.

Methods of Compliance

To the extent feasible, engineering and work practice controls must be used to reduce employee exposure to below the PEL and/or EL. Regardless of exposure levels, the following control methods must be used for all activities:

- HEPA vacuums to collect debris and dust;
- wet methods; and
- prompt cleanup and disposal of waste and debris.

In addition to the methods above, the following control methods must be used to comply with the PEL/EL:

- local exhaust ventilation,
- enclosure or isolation of the work process,
- ventilation of the regulated area,
- other feasible work practices or engineering controls.

Work practices that are prohibited include:

- the use of high speed abrasive disk saws,
- compressed air,
- dry sweeping,
- employee rotation to reduce exposures.

<u>Class I Requirements</u>:

In addition to the work practices and control methods presented above, the following methods of compliance shall be used:

- all work must be supervised by a competent person;
- isolate the heating, ventilation, and cooling system;
- use impermeable drop cloths beneath all removal activity;
- cover all nonmovable objects
- ventilate the regulated area to move dust away from the employee;
- use critical barriers or other isolation system in combination with perimeter air monitoring if:
  - ✓ the job involves >25 linear feet or >10 square feet of TSI/surfacing material;
  - ✓ where a negative exposure assessment has not been done, or
  - ✓ where employees are working adjacent to Class I activity.

One or more of the following specific control methods must be utilized when conducting Class I asbestos activities:

- negative pressure enclosures
- glove bags
- negative pressure glove bag
- negative pressure glove box
- water spray process
- small, walk-in enclosures
- alternate control method certified by a CIH or PE who is a project designer.

<u>Class II Requirements</u>:

- competent person supervision
- indoor removals without a negative exposure assessment must:
  - -use critical barriers
  - -other feasible barrier/isolation methods
  - -listed work practices for each type of work
  - -employees must be trained and use work practices/controls specifically outlined in the standard based on the type of material involved (i.e., flooring, roofing, gaskets, etc.)

NOTE: Protective clothing and respirators are not required until the PEL levels are met EXCEPT K-12 school projects unless O & M.

<u>Class III Requirements</u>:

- minimize exposure to individual performing work and bystanders
- use local exhaust when feasible
- use mini-enclosures or glovebag systems when cutting, drilling, abrading, sanding, chipping, breaking, or sawing TSI and/or surfacing material
- if exposures are above the PEL/EL or a negative exposure assessment was not performed:
  - contain area using drop cloths and plastic barriers, <u>or</u>
  - isolate area using negative pressure enclosure, glovebag, etc

  Use respirator if:
  - disturbing TSI or surfacing material
  - PEL/EL exceeded
  - a negative exposure assessment is not available

Class IV Requirements

Work practices required include:

- use of wet methods
- use of HEPA vacuums
- prompt action to clean up ACM/PACM debris
- following paragraph H respirator requirements
- assume waste and debris in areas with accessible (friable) TSI and surfacing materials contains asbestos

Respiratory Protection

An employer must provide respirators and ensure they are used:

- during all Class I work;
- during all Class II work where ACM is not removed substantially intact;
- during all Class II and III work that is not performed wet;
- during all Class II and III work where TSI or surfacing ACM/PACM is disturbed;
- during all Class IV work within a regulated area where other employees are required to wear respirators;
- during all work where exposures exceed the PEL or EL; and
- in emergencies.

Respirators must be selected according to the provisions of 30CFR Part 11. The employers must develop a respiratory program in accordance with the Respirator Standard for General Industry (29 CFR 1910.134).

Employees who use a filter respirator must change filters whenever an increase in breathing resistance is detected. Employees who wear respirators must be allowed to wash their face and respirator face piece whenever necessary to prevent skin irritation associated with respirator use. An employee must not be assigned to tasks requiring the use of respirators if a physician determines that the employee is unable to function normally while wearing a respirator or that the employee's safety and health or that of others would be affected by the employee's use of a respirator. In this case, the employer must assign the employee to another job or give the employee the opportunity to transfer to a different job that does not require the use of a respirator. The job should be with the same employer, in the same geographical area, and with the same seniority, status, and rate of pay, if such a position is available.

Employers must assure that a respirator issued to an employee fits properly and exhibits minimum face piece leakage. Employers must perform quantitative or qualitative fit tests at the time of initial fitting and at least every twelve months for each employee wearing negative-pressure respirators.

Protective Clothing

An employer must provide and require the use of protective clothing such as coveralls or similar full-body clothing, head coverings, gloves, and foot coverings: for any employee exposed to airborne concentrations of asbestos that exceed the PEL and/or EL;

- in situations where a negative exposure assessment was required and not performed; and
- for Class I work involving TSI or surfacing material in amounts exceeding 25 linear feet or 10 square feet.

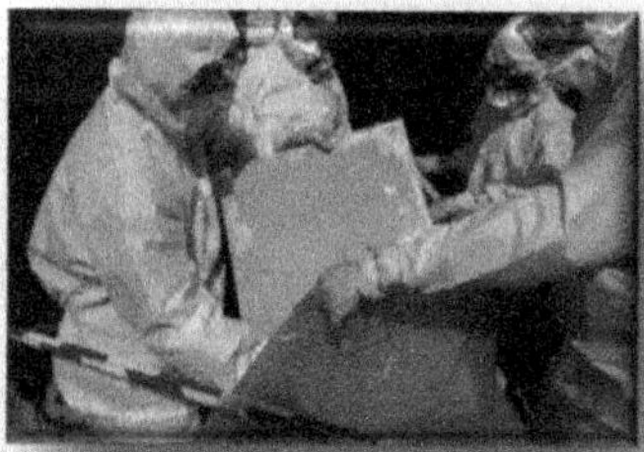

Wherever the possibility of eye irritation exists, face shields, vented goggles, or other appropriate protective equipment must be provided and worn. Asbestos-contaminated work clothing must be removed in change rooms and placed and stored in closed, labeled containers that prevent dispersion of the asbestos into the ambient environment. Protective clothing and equipment must be cleaned, laundered, repaired, or replaced to maintain their effectiveness. The employer must inform any person who launders or cleans asbestos- contaminated clothing or equipment of the potentially harmful effects of exposure to asbestos. Contaminated clothing and equipment taken out of change rooms or the workplace for cleaning, maintenance or disposal must be transported in sealed impermeable bags, or other closed impermeable containers and be appropriately labeled.

Hygiene Facilities and Practices

The requirements for hygiene facilities and practices differ based on the class of work activity involved.

Class I Work (>25 linear feet or 10 square feet)

- the employer must provide clean change areas equipped with separate storage facilities for protective clothing and street clothing;
- the employer must provide lunch areas in which the airborne concentrations of asbestos are below the PEL/EL;
- the employer must establish a decontamination area for the decontamination of asbestos-contaminated employees that is adjacent and connected to the regulated area. The decontamination area must consist of an equipment room, shower, area, and clean room in series.

Clean rooms must be equipped with a locker or appropriate storage container for each employee. Equipment rooms must be supplied with impermeable, labeled bags and containers for the containment and disposal of asbestos-contaminated protective clothing and equipment. Where feasible, shower facilities must be contiguous to the equipment room and the clean change room. When contiguous decontamination facilities are not feasible, the workers must remore contamination from suits or put on clean protective clothing, then proceed to the shower. Employers must ensure that employees enter and exit the regulated area through the decontamination area.

Class I, II, and III work

The following hygiene facility and practice requirements apply to Class I work involving less than 25 linear feet or 10 square feet of material, and Class II and III work if exposures exceed the PEL or if a negative exposure assessment has not been done:

- establish equipment room adjacent to the regulated area;
- all equipment/containers and clothing must be cleaned prior to removal from the equipment room;
- employees must enter and exit through the equipment room.

Class IV Work

- employees cleaning up TSI or surfacing materials must be provided an area to decontaminate equipment and personnel;
- employees working in a requlated area must meet hygiene practices requirements of the regulated area.

An employer shall ensure that employees do not smoke in work areas where they are occupationally exposed to asbestos because of activities in that work area.

Information and Training

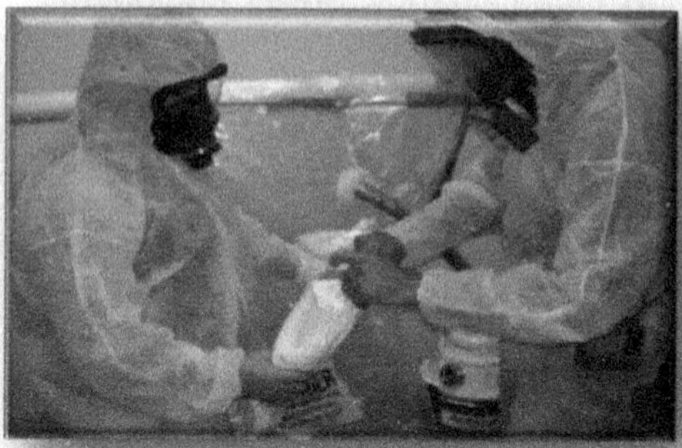

An employer must provide training, at no cost to the employee, for all employees who install asbestos- containing products, or who perform Class I, II, III, or IV asbestos operations. Training must be provided prior to or at the time of initial assignment and at least yearly thereafter. OSHA bases its training requirements on the type of activity being conducted.

- training for Class I operations must be the equivalent of the 4-day asbestos abatement worker training outlined in the EPA Model Accreditation Plan;
- training for Class II operations which involve asbestos-containing roofing materials, flooring materials, siding materials, ceiling tiles, or transite panels must include training topics outlined by OSHA, including hands-on, for a minimum of 8 hours. Other Class II activity training must cover OSHA-required topics, include hand-on training, and cover the OSHA work practices and engineering control requirements detailed for Class II work.
- training for Class III operations must be the equivalent of the 16-hour maintenance and custodial training detailed in the AHERA regulations;
- training for Class IV operations must be equivalent to the requirements for EPA 2-hour awareness training outlined in AHERA.

NOTE : Your instructor will provide you
with local and state training requirements.

All training programs must inform employees about the

- methods of recognizing asbestos,
- health hazards of asbestos exposure,
- relationship between asbestos and smoking in producing lung cancer,
- operations that could result in asbestos exposure,

- importance of necessary protective controls to minimize exposure including, as applicable, engineering controls, work practices, respirators, housekeeping procedures, emergency procedures, and waste disposal procedures,
- purpose, proper use and limitations of respirators,
- the medical surveillance program,
- a complete review of the OSHA standard(s) including appendices.

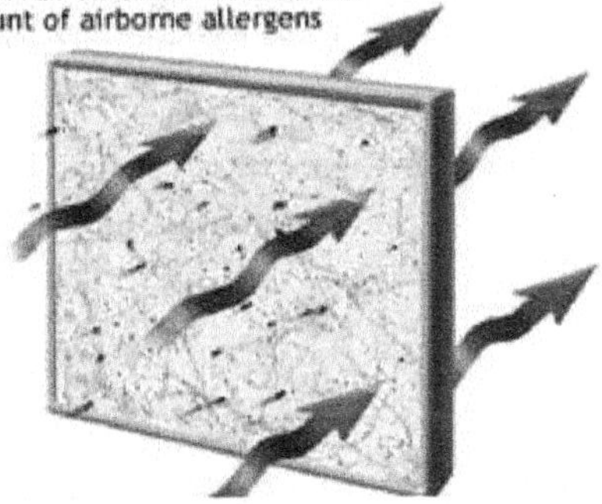
A HEPA air filter can reduce the amount of airborne allergens

The employer is also required to inform all employees concerning the availability of self-help smoking cessation program material and to provide the names, addresses and phone numbers of public health organizations that provide information, materials and/or conduct programs concerning smoking cessation. Appendix J of the construction standard, which contains a list of such organizations, may be used to comply with this requirement. The training must also cover the requirements for posting signs and affixing labels and the meaning of the required legends for such signs and labels.

If an employer's Class II work involves only the removal and/or disturbance of one generic category of building material, for example roofing materials, flooring materials, siding, etc., the employer is required to provide training that includes all the information listed above, including a "hands-on" component, and be at least eight hours in length.

All training materials must be available to the employees without cost and, upon request, to the Assistant Secretary for OSHA and the Director of the National Institute for Occupational Safety and Health (NIOSH).

Housekeeping

Vacuuming equipment, when used, must have HEPA filters. Dust and debris in areas with accessible TSI, surfacing material or visibly deteriorated ACM must be HEPA vacuumed and **not** dry swept. Asbestos waste, scrap, debris, bags, containers, equipment, and asbestos-contaminated clothing consigned for disposal must be collected and disposed of in sealed, labeled, impermeable bags or other closed, labeled impermeable containers.

Medical Surveillance

The employer must establish a medical surveillance program prior to assignment, for all employees who 1) will be required to wear negative-pressure respirators, 2) will be engaged in Class I, II, and III work for 30 or more days per year, or 3) will be exposed to airborne concentrations of asbestos at or above the PEL and/or EL. (OSHA medical screening is required prior to the use of any respirator),

All examinations must be performed under the supervision of a licensed physician and shall be provided without cost to the employee and at a reasonable time and place. Examinations must include:

- a medical and work history;
- a physical examination with special emphasis directed to the respiratory, cardiovascular, and gastrointestinal systems;
- completion of a respiratory disease questionnaire;
- a pulmonary function test.

A chest X-ray may be administered at the discretion of the physician. These examinations must be made available annually or more often if the physician deems it necessary. The employer must give the examining physician a copy of the standard and its Appendices D, E, G, and I; a description of the employee's duties relating to the employee's asbestos exposure; the exposure level or anticipated exposure level; a description of any personal protective and respiratory equipment used or to be used; and information from previous medical examinations.

Also, employers must obtain a written signed opinion from the physician that contains: the results of the medical examination and the physician's opinion as to whether the employee has any detected medical conditions that would place the employee at an increased risk from exposure to asbestos, any recommended limitations on the employee or upon the use of personal protective equipment such as clothing or respirators, and statements that the employee has been informed by the physician of the increased risk of lung cancer attributable to the combined effects of smoking and working with asbestos and the results of the medical examination.

The physician is not to reveal in the written opinion given to the employer specific findings or diagnoses unrelated to occupational exposure to asbestos. Finally, the employer must provide a copy of the physician's written opinion to the affected employee within 30 days from its receipt.

Recordkeeping

Employers must keep an accurate record of all measurements taken to monitor employee exposure to asbestos. This record should include the date of measurement, operation or activity involving exposure, sampling and analytical methods used and evidence of their accuracy; number, duration, and results of samples taken; type of respiratory protective devices worn; and name, social security number, and the results of all employee exposure measurements. These records must be kept for 30 years.

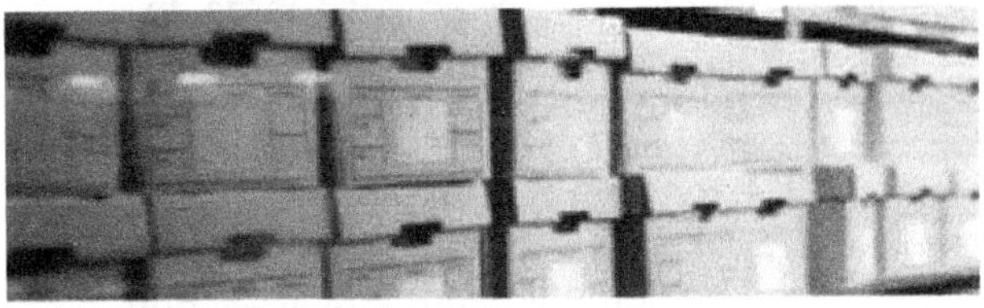

Employers who have relied on objective data as a negative exposure assessment must establish and maintain an accurate record of this data in support of the monitoring exemption. This data must include the product qualifying for the exemption; the source of the data; the testing protocol, results of testing and/or analysis of the material for release of asbestos; a description of the operation exempted and how the data support the exemption; and other data relevant to the operations materials, processing, or employee exposures covered by the exemption. These records shall be maintained for as long as the employer relies on the data for exemptions.

Likewise, the employer must maintain an accurate record for each employee subject to medical surveillance. The record must include: the name and social security number of the employee; a copy of the employee's medical examination results; physician's written opinions; any employee medical complaints related to exposure to asbestos; and information provided to the examining physician as described under medical surveillance. This record must be maintained for the duration of employment plus 30 years.

The employer must maintain all employee training records for one year beyond the last date of employment by that employee.

All records must be made available on request to the Assistant Secretary for OSHA, the Director of the National Institute for Occupational Safety and Health (NIOSH), affected employees, former employees, and designated representatives. When the employer ceases to do business and there is no successor employer to receive the records for the prescribed period, the employer must notify the Director of NIOSH at least 90 days prior to disposal of records.

U.S. ENVIRONMENTAL PROTECTION AGENCY ASBESTOS REGULATIONS

EPA WORKER PROTECTION RULE

This regulation extends the OSHA standards to state and local employees who perform asbestos work and who are not covered by the OSHA Asbestos Standards, or by a state OSHA plan. The Rule currently parallels 1986 OSHA requirements and covers medical examinations, air monitoring and reporting, protective equipment, work practices, and recordkeeping. Currently, the EPA Worker Protection Rule does not include the excursion limit requirements found in the OSHA standards.

NATIONAL EMISSION STANDARDS FOR HAZARDOUS AIR POLLUTANTS (NESHAP)

EPA's rules concerning the application, removal, and disposal of asbestos- containing materials were issued under NESHAP. Also included in NESHAP are rules concerning manufacturing, spraying, and fabricating of asbestos-containing material. NESHAP was revised 20 November 1990 to clarify requirements regarding removal and disposal of asbestos-containing materials.

Bans on Asbestos-containing Material

Three bans on asbestos-containing material were set forth by the NESHAP regulations. These bans occurred in the three years as indicated below:

1973 - Spray-applied insulating materials
1976 - Pre-molded insulation, if friable
1978 - Spray-applied decorative material

NESHAP Definitions (Selected)

Category I Nonfriable ACM includes asbestos-containing packing, gaskets, resilient floor coverings and asphalt roofing products containing more than 1% asbestos. (Category I nonfriable ACM has been interpreted to include pliable asbestos-containing sealants and mastics since they exhibit many of the same characteristics of Category I nonfriable asbestos-containing material.

Category II Nonfriable ACM includes any material, excluding Category I nonfriable ACM, containing more than 1% asbestos that, when dry cannot be crumbled, pulverized, or reduced to powder by hand pressure.

Friable asbestos material includes any material containing more than 1% asbestos that, when dry, can be crumbled, pulverized, or reduced to powder by hand pressure.

Regulated asbestos-containing material (RACM) includes

- friable ACM;
- Category I nonfriable ACM that has become friable;
- Category I nonfriable ACM that will be or has been subjected to sanding, grinding, cutting, or abrading;
- Category II nonfriable ACM that has a high probability of becoming or has become crumbled, pulverized, or reduced to powder by the forces expected to act on the material in the
- course of demolition or renovation operations regulated by NESHAP.

Waste Generator includes any owner or operator of a source whose act or process produces asbestos-containing waste material.

Analytical Methods

Polarized light microscopy (PLM) analysis of bulk samples is required by NESHAP to determine the asbestos content of ACM. If the asbestos content is less than 10% as determined by a method other than point counting by PLM, verify the asbestos content by point counting using PLM. (See NESHAP SUMMARY for more information regarding the use of point counting.)

Notification

Specific notification to a regional or state NESHAP Coordinator is required before a building is demolished or renovated. This written notification must be delivered by U.S. Postal Service, commercial delivery service, or hand delivery. Some states have notification requirements for removal of RACM in amounts less than federal NESHAP, and/or regulate nonfriable materials. Check with the state agency responsible for NESHAP notification before any project begins.

NOTE: LOCAL AND STATE REQUIREMENTS MAY DIFFER.

NESHAP SUMMARY

DEMOLITION			RENOVATION	
AMOUNT OF RACM	>260 if 160 ft^2 OR 35 ft^3	<260 ft^2 if , 160 ft^2 or 35 ft^3	> 260 lf, 160 ft^2 or 35 ft^3	>260 lf, 160 ft^2 or 35 ft^3
NOTIFICATION	Yes all requirements apply	Yes Simple Notification	Yes all requirements apply	Consider yearly (additive) amounts
HOW FAR IN ADVANCE?	10 days	10 days	10 days	At least 10 working days prior to end of calendar year*

DEMOLITON BY ORDER

Notification: Same as for demolition*

How far in advance: Answer: early as possible, but not later than the following day**

*Also include name, title & authority of government representative, date order issued and date which demolition was ordered to begin.
A copy of the order shall be attached to the notification.

** Update notice, as necessary, including when the amount of asbestos affected changes by at least 20 percent. A working date includes Monday through Friday and holidays which fall on any of the days Monday through Friday.

NESHAP requires the removal of all RACM from a facility being demolished or renovated before any activity begins that would break up, dislodge, or similarly disturb the material or preclude access to the material for subsequent removal. The material must be removed using wet removal techniques. (With special approval from EPA, dry removal is allowed under certain circumstances.) No visible emissions to the outside air are permitted during removal or renovation. If a facility is demolished by intentional burning, all RACM including Category I and Category II nonfriable ACM must be removed in accordance with NESHAP before burning.

The following information is required on the notification:

(NOTE: In a facility being demolished, if the combined amount of RACM is less than 260 lf, 160 sf, or 35 cf, only simple notification requirements apply. This notification must include the information in **bold typeface**, with the 10-day notification requirement also applying.)

- An indication of whether the notice is the original or a revised notification;
- Name, address and phone number of both the facility owner and operator and the asbestos removal contractor owner or operator;
- Type of operation: demolition or renovation;
- Description of the facility or affected part of the facility including the size (square feet and number of floors), age and present and prior use of the facility;
- Procedures, including analytical methods, employed to detect the presence of RACM and Category I and Category II nonfriable ACM;
- Estimate of the approximate amount of RACM to be removed from the facility; also estimate the approximate amount of Category I and Category II nonfriable ACM in the affected part of the facility that will not be removed before demolition;
- Location and street address (including building number or name and floor or room numbers, if appropriate), city, county, and state of the facility being demolished or renovated;
- Scheduled starting and completion dates for asbestos removal work (or any other activity, such as site preparation that would break up, dislodge, or similarly disturb asbestos material) in a demolition or renovation; planned renovation operations involving individual nonscheduled operations shall only include the beginning and ending dates of the report period;
- Scheduled starting and completion dates of demolition or renovation;
- Description of planned demolition and renovation work to be performed and methods;
- Description of work practices and engineering controls to be used to comply with the requirements of NESHAP, including asbestos removal and waste-handling emission control procedures;
- Name and location of the waste disposal site where the asbestos-containing waste material will be deposited;
- A certification that at least one on-site representative (such as a foreman or management level person) trained in the asbestos demolition or renovation provisions and the means of complying with them, be present when RACM is stripped, removed, or otherwise handled or disturbed as described by

this notification;

- If demolition is under order form a state or local agency, the name, title, and authority of the state or local government representative who ordered the demolition, the date that the order was issued, and the date on which the demolition was ordered to begin. A copy of the order shall be attached to the notification;
- For emergency renovations, the date and hour that the emergency occurred, a description of the sudden, unexpected event, and an explanation of how the event caused an unsafe condition, or would cause equipment damage or an unreasonable financial burden;
- Description of procedures to be followed in the event that unexpected RACM is found or Category II nonfriable ACM becomes crumbed, pulverized, or reduced to powder;
- Name, address, and telephone number of the transporter.

Below are listed the exemptions from removal according to

NESHAP:

- It is Category I nonfriable ACM that is not in poor condition and is not friable.
- It is on a facility component that is encased in concrete or other similarly hard material and is adequately wet whenever exposed during demolition.
- It was not accessible for testing and was, therefore, not discovered until after demolition began and as a result of the demolition, the material cannot be safely removed. If not removed for safety reasons, the exposed RACM and any asbestos-containing debris must be treated as asbestos- containing waste material and adequately wet at all times until disposed of.
- The material is Category II nonfriable ACM and the probability is low that the materials will become crumbled, pulverized, or reduced to powder during demolition.

NOTE: EPA has determined that any demolition operation (i.e., use of a wrecking ball, implosion, use of a bulldozer, backhoe, or other heavy machinery to knock a building over) will extensively damage Category II ACM such that it is crumbled, pulverized, or reduced to powder.

Waste Disposal

Below are listed the waste disposal provisions required under NESHAP. (For

more information, see the *Waste Disposal* section of this manual.)

- No visible emissions to the outside air are allowed during collection, packaging, transportation, or deposition of ACM waste.
- Wet ACM must be sealed in a leak-tight container.
- Containers must be labeled with OSHA warning labels.
- For waste that will be transported off the site of the facility, label all containers or wrapped materials with the name of the waste generator and the location at which the waste was generated.
- Mark vehicles used to transport asbestos-containing waste material during the loading and unloading of waste so that the signs are visible.
- Maintain waste shipment records using a form similar to Figure 4 included in the NESHAP regulations (see Waste Disposal Form in the *Waste Disposal* section). A copy of the waste shipment record must be provided to the disposal site owners or operators at the same time as the asbestos-containing waste material is delivered to the disposal site. A copy of the waste shipment record, signed by the owner or operator of the designated disposal site must be returned to the waste generator.

Since NESHAP mandates the removal of friable ACM before a building is demolished or renovated, if the renovation will disturb the ACM, any plan for managing ACM should take into account the costs of eventual removal. Certain abatement methods such as encapsulation and enclosure may make eventual removal more difficult.

ASBESTOS HAZARD EMERGENCY RESPONSE ACT

In October 1986, the Asbestos Hazard Emergency Response Act (AHERA) was signed into law. Included in this Act were provisions directing the EPA to establish rules and regulations addressing asbestos-containing materials in schools. Specifically, EPA was directed to address the issues of: (1) identifying, (2) evaluating, and (3) controlling ACM in schools.

Under a six-month deadline set by Congress, EPA convened a panel of representatives from various associations and interest groups that would be affected by the regulation. The panel members negotiated on many issues, and the results were published during April of 1987 in the form of a proposed rule. Final AHERA regulations (rules) became effective 14 December 1987.

They are found in 40 CFR 763 Subpart E sections §763.80 - §763.99 under the Toxic Substances Control Act (TSCA).

The regulations require that all public and private elementary and secondary schools (K-12) inspect for both friable and nonfriable asbestos, implement response actions, and submit asbestos management plans to state governors or designated agencies. Schools were given until 9 May 1989 to submit the required management plans to the state governors or designated agencies. States had 90 days to approve or disapprove each plan submitted. Management plans had to be implemented by 9 July 1989, and be completed in a timely fashion.

Schools must use accredited persons to:

- conduct inspections,
- develop management plans, and
- design or conduct response actions.

Also, the rule requires periodic surveillance and re-inspection to monitor asbestos-containing materials left in schools. Periodic surveillance requires checking these materials every six months to determine if there has been a change in condition of the material since the last inspection or surveillance. In addition, schools must have an accredited inspector re-inspect and reassess the condition of remaining asbestos-containing materials every three years and determine if the condition of the material requires new response action activity.

Schools that fail to conduct the inspections, knowingly submit false evidence to their governors, or fail to develop a management plan in accordance with regulations, can be assess a civil penalty under the Toxic Substances Control Act (TSCA) of up to $5,000 for each day the school is in violation. AHERA also provides that civil penalties assessed will be used by schools to comply with AHERA requirements. Unspent portions of the assessed civil penalties will be deposited in a federal Asbestos Trust Fund. These monies will be made available for further asbestos abatement activities.

Schools that had previously conducted inspections consistent with this final rule and had determined that no asbestos-containing material is present in the schools were excluded from the inspection requirements. In addition, a school is exempt if it was built after 12 October 1988, and an architect, project engineer or accredited inspector signed a statement that no asbestos-containing material had been specified for use in construction documents. States may receive a waiver from some or all of the requirements of the proposed rule if they have established and are implementing (or intend to implement) a program of asbestos inspection and management at least as stringent as the requirements of the final rule.

Your Instruction will provide you with any local, state or more current legislation and regulations.

ASBESTOS: MANUFACTURE, IMPORTATION, PROCESSING AND DISTRIBUTION IN COMMERCE PROHIBITIONS; FINAL RULE (BAN AND PHASE-OUT RULE)

On July 7, 1989, EPA announced the promulgation of its long-awaited asbestos ban and phase-down rule. This rule, which was to be phased in over a seven-year period beginning in 1990, prohibited the manufacture, importation, processing, and distribution of certain commercially available asbestos-containing products. This rule would have effectively banned the use of nearly 95% of all asbestos products used in the United States, with the exception of products for which no acceptable substitute has been found, and certain products for military use.

EPA had adopted separate dates for the banning of the manufacture, importation, and processing of asbestos-containing products, and for the distribution of asbestos-containing products in commerce. However, this regulation was vacated by the Fifth Circuit Court of Appeals in October, 1991. EPA appealed the court's decision and the appeal was rejected on 27 November,1991.

The court did allow EPA to ban new uses of certain asbestos-containing products and those products that were not being manufactured, imported, or processed on the date the final rule was issued (12 July 1989). EPA issued a notice in the *Federal Register* that requested information on the status of 14 product categories included in the rule that were not being manufactured, processed, or imported when the final rule was published. Based on the research conducted by EPA, and information provided by commenters, EPA published in the 5 November 1993 *Federal Register* the following six asbestos-containing products that are still subject to the Ban and Phase-out Rule:

- corrugated paper
- rollboard
- commercial paper
- specialty paper
- flooring felt, and
- new uses of asbestos.

Your Instruction will provide you with any local, state or more current legislation and regulations.

ASBESTOS SCHOOL HAZARD ABATEMENT REAUTHORIZATION ACT (ASHARA) [covering Public and Commercial Buildings]

ASHARA Regulations cover Public and Commercial Buildings

Section 206 of the Toxic Substances Control Act (TSCA) mandated that EPA develop an asbestos Model Accreditation Plan (MAP). The original MAP was promulgated in 1987 and became codified as 40 CFR Part 763, Appendix C to Subpart E. Section 206 of TSCA was later amended by the Asbestos School Hazard Abatement Reauthorization Act (ASHARA). ASHARA mandated that the MAP be revised to:

- provide for the extension of accreditation requirements to public and commercial buildings for persons who inspect for asbestos-containing material, design response actions, or carry out response actions; and
- increase the minimum number of training hours, including additional hands-on training, required for accreditation of workers and supervisors performing work in schools and/or public and commercial buildings.

ASHARA does not require persons who prepare management plans in public or commercial buildings to obtain accreditation. The accreditation requirements of the ASHARA statute went into effect on 28 November 1992. The revised MAP, which provides more information the meaning of the new statutory requirements and expands the length of, and/or topics addressed in the training courses, were published as an interim final rule in the *Federal Register* on 3 February 1994, and took effect on 4 April 1994.

ASHARA requires that accredited asbestos abatement contractors/supervisors and accredited workers are used to supervise or carry out:

- response actions other than a small-scale, short duration activity,
- maintenance activities that disturb friable asbestos-containing material other than a small-scale, short duration activity,
- a response action for a major fiber release episode (the accidental or unintentional falling or dislodging of greater than 3 square or 3 linear feet of ACM.)

STATE AND LOCAL REGULATIONS

Several provisions in AHERA and ASHARA encourage states to develop their own regulatory programs. For example, states are encouraged to establish and operate training and certification programs for the various categories of asbestos professionals, as long as the programs are at least as stringent as AHERA's Model Plan. In addition, some states have established requirements that exceed EPA's in the area of notification of abatement actions, abatement work practices, and transportation and disposal of asbestos-contaminated waste. Inspectors/management planners should consult state and local regulatory agencies in their areas.

The following exhibit outlines federal agency regulations both by year and by regulatory action. Exhibit 1 Chronology of Major Federal Initiatives

Your Instruction will provide you with any local, state or more current legislation and regulations.

Exhibit 1 Chronology of Major Federal Initiatives

YEAR	EPA	OSHA	OTHER
1971	Asbestos listed as hazardous Air pollutant.	Existing occupational exposure	
1973	Standard for milling, manufacturing and building demolition. Spray application of friable material (>1%) prohibited		
1973	No visible emissions.		
1974	Effluent guidelines for manufacturing sources (EPA/FWPCA)		
1975	No visible emissions standard Extended to waste collection, disposal and processing industries not previously covered.		
1976		Occupational exposure standard lowered to 2	
1977			Consumer Product Safety Commission prohibition of asbestos in patching compounds and emberizing agents.
1978	All friable spray-on material prohibited, all demolition and renovation covered by no visible emissions standard (EPA/NESHAP)		Controls regarding transport of friable ACM (DOT)
1979	Technical assistance program to schools initiated to identify and control friable ACM		Controls regarding transport of friable ACM (DOT)
1982	Identification and notification of friable ACM in schools rule (EPA/TSCA).		
	Required reporting of production and exposure data on asbestos (EPA/TSCA).		
1984	EPA/NESHAP standard formally recognized		
	Asbestos School Hazard Abatement Act (ASHAA) – loan and grant program to help eliminate hazards		
1986	AHERA	Construction Industry Standard issued. Permissible exposure level lowered	
1987	TSCA amended to reflect AHERA		
	Worker Protection Rule		
1988		Amendment of general and construction industry asbestos	
1989	Ban & Phase-Out Rule		

1990	Extensive NESHAP revisions including Category I & II nonfriable material definitions, point counting and waste disposal manifests.	Court-ordered amendments to asbestos standard for the construction industry regarding informing employees of the	DOT Hazardous Materials Regulations (HMR) 49 CFR Part 107 et. al. published based on UN standards. Included new asbestos classification, hazard communication, packaging and handling requirements.
1991	Fifth Circuit Court of Appeals vacates most of Ban & Phase-out Rule		
1992	ASHAA reauthorized (ASHARA), requiring accreditation of designers, inspectors, contractors/supervisors and workers involved in asbestos detection and remediation in public and commercial buildings.		DOT HRM revised with less stringent requirements when shipping friable asbestos within the
	EPA issues draft list of asbestos products still covered by Ban and Phase-Out Rule.		
1993	EPA published list of asbestos-containing products which are banned from manufacture, importation, processing and distribution.		
1994	ASHARA Interim Final Rule published including revisions to MAP.	Revisions to General and Construction Industry Standards and Issuance of Shipyard Standard	
1996		Interpretative letters clarify bulk sample analysis of drywall or gypsum board, joint compound, & tape be performed separately (where feasible), if any of the layers contain>1% worker protection is required.	
1998		Interpretative letters clarify that bulk sampling procedures are considered class III operations and hard wall/ceiling plasters and stucco are not considered surfacing materials	
1999		Respiratory Protection Standard revised to require fit testing for all tight fitting masks and to be performed at least annually, the assigned protection factors for qualitative fit tests dropped to10 x's, and the approval of mobile quantitative fit testing devices.	
2000	Worker Protection Rule amended to incorporate both Construction & General Industry Asbestos Standards		
2008		Respiratory Protection Standard revised to only allow qualitative fit testing on respirators requiring a fit factor of 100 or less.	

Acoustical Insulation -The general application or use of asbestos for the control of sound due to its lack of reverberant surfaces.

Acoustical Tile - A finishing material in a building usually found in the ceiling or walls for the purpose of noise control.

Actinolite - One of six naturally occurring asbestos minerals. It is not normally used commercially.

Addenda - Changes made to working drawings and specifications for a building before the work is bid.

Aggressive Sampling - Air sampling which takes place after final clean-up while the air is being physically agitated to produce a "worst case" situation.

AHERA - Asbestos Hazard Emergency Response Act of 1986
In 1986, the Asbestos Hazard Emergency Response Act (AHERA) was signed into law as Title II of the Toxic Substance Control Act. Additionally, the **Asbestos School Hazard Abatement Reauthorization Act** (ASHARA), passed in 1990, requires accreditation of personnel working on asbestos activities in schools, and public and commercial buildings.

AIHA Accredited Laboratory –A certification given by the AIHA to an analytical laboratory that has successfully participated in the "Proficiency Analytical Testing" program for quality control as established by the National Institute for Occupational Safety and Health.

Air Lock – A system of enclosures consisting of two polyethylene curtained doorways at least three feet apart that does not permit air movement between clean and contaminated areas.

Air Monitoring – The process of measuring the airborne fiber concentration of a specific quantity of air over a given amount of time.

Air Plenum – Any space used to convey air in a building or structure. The space above a suspended ceiling is often used as an air plenum.

Air-purifying respirator means a respirator with an air-purifying filter, cartridge, or canister that removes specific air contaminants by passing ambient air through the air-purifying element.

Alveoli - Located in clusters around the respiratory bronchioles of the lungs, this is the area in which true respiration takes place.

Amended Water – Water to which a chemical wetting agent (surfactant) has been added to improve penetration into ACM that is being removed.

Amosite - An asbestiform mineral of the amphibole group. It is the second most commonly used form of asbestos in the U.S. Also known as brown asbestos.

Amphibole - One of the two major groups of minerals from which the asbestiform minerals are derived, distinguished by their chain-like crystal structure and chemical composition. Amosite and crocidolite are examples of amphibole minerals.

Anthophyllite - One of six naturally occurring asbestos minerals. It is of limited commercial value.

Asbestos - A generic name given to a number of naturally occurring hydrated mineral silicates that possess a unique crystalline structure, are incombustible in air, and separate into fibers. Asbestos includes the asbestiform varieties of chrysotile (serpentine); crocidolite (riebeckite); amosite (cummingtonite-

grunerite); anthophyllite; tremolite, and actinolite.
Asbestos-Containing Material (ACBM) - Surfacing ACM, thermal system insulation ACM, or miscellaneous ACM that is found in or on interstructural members or other parts of a school building (AHERA definition).
Asbestos-Containing Material - Any material or product which contains more than 1 percent asbestos (AHERA, OSHA definition).
Asbestos Fibers – Fibers with their length being greater than five microns (length to width ratio of 3:1); generated from an asbestos-containing material.
Asbestosis - A non-malignant, progressive, irreversible lung disease caused by the inhalation of asbestos dust and characterized by diffuse fibrosis.
ASHARA - Asbestos School Hazard Abatement Reauthorization Act. U.S. EPA sets forth accreditation requirements for inspectors, contractor, supervisors, designers, and workers to public and commercial buildings.
Assigned protection factor (APF) means the workplace level of respiratory protection that a respirator or class of respirators is expected to provide to employees when the employer implements a continuing, effective respiratory protection program as specified.
Authorized person means any person authorized by the employer and required by work duties to be present in regulated areas. Building/facility owner is the legal entity, including a lessee, which exercises control over management and record keeping functions relating to a building and/or facility in which activities covered by this standard take place.
Breeching- A duct which transports combustion gases from a boiler or heater to a chimney or stack. Also called a flue.
Bridging Encapsulant – The application of a non-penetrating sealant over the surface of asbestos-containing material to prevent the release of asbestos fibers.
Building Inspector - A person who conducts a survey of a building for the presence of asbestos-containing materials. Must be accredited under AHERA and ASHARA regulations.
Building/facility *owner* is the legal entity, including a lessee, which exercises control over management and record keeping functions relating to a building and/or facility in which activities covered by this standard take place.
Bulk Sample - A sample of suspected media (e.g., soil or dust) is obtained from a site to be analyzed microscopically for asbestos content. Bulk sample analysis can be part of a process to assess the hazard from asbestos at a site.
Canister or cartridge means a container with a filter, sorbent, or catalyst, or combination of these items, which removes specific contaminants from the air passed through the container.
Category I – Non-friable ACM - Asbestos-containing packings, gaskets, resilient floor coverings, and asphalt roofing products containing more than 1% asbestos, which are not capable of being crumbled, pulverized, or reduced to powder by hand pressure when dry. Defined by NESHAP regulations.

Category II – Non-friable ACM - All asbestos-containing materials not defined as Category I Nonfriable ACM, containing more than 1% asbestos, which cannot be crumbled, pulverized, or reduced to powder by hand pressure when dry. Defined by the NESHAP regulations. Example: asbestos cement products.

Cementitious ACM - Asbestos-containing materials that are densely packed, granular and are generally non-friable.

Certified Industrial Hygienist (CIH) means one certified in the practice of industrial hygiene by the American Board of Industrial Hygiene.

Chain-of custody - Formal procedures for tracking samples and insuring their integrity.

Change Order - A change made to construction documents after a contract for construction has been signed.

Chrysotile - The only asbestiform mineral of the serpentine group. It is the most common form of asbestos used in buildings. Also known as white asbestos.

Certified Industrial Hygienist - An industrial hygienist who has been granted certification by the American Board of Industrial Hygiene.

CFR – Code of Federal Regulations

Cilia - Tiny hair-like structures in the windpipe and bronchi of the lung passages which beat upward and that help force undesirable particles, fibers and liquids up and out of the lungs.

Claims-Made Insurance - Insurance in which a claim is allowed only if the insurance is in effect when the claim is made, that is, when the injury or effect is observed.

Class I Asbestos Activity - Activities involving the removal of thermal system insulation and surfacing ACM and PACM.

Class II Asbestos Activity - Activities involving the removal of ACM which is not TSI or surfacing material. From OSHA Construction Industry Standards.

Class III Asbestos Activity - Repair and maintenance operations where ACM is likely to be disturbed. From OSHA Construction Industry Standards.

Class IV Asbestos Activity - Maintenance and custodial construction activities during which employees contact but do not disturb ACM or PACM, and activities to clean up dust, waste and debris resulting from Class I, II, and III activities.From OSHA Construction Industry Standards.

Clean Room - The first stage of the decontamination enclosure system in which workers prepare to enter the work area by removing street clothes and putting on protective clothing and respirators.

Clearance Sampling - Air samples collected at the conclusion of an asbestos response action to determine if airborne asbestos fiber concentrations are below those levels acceptable for persons to reoccupy an area.

Competent Person - Defined by the OSHA Construction Industry Asbestos Standards as a person who is capable of identifying existing hazards in the workplace and selecting the appropriate control strategy for asbestos exposure, and who has the authority to take prompt corrective measures to eliminate them.

Contaminated Items – Any objects that have been exposed to airborne asbestos fibers without being sealed off or isolated.

Contract Documents - Legally binding building drawings and specifications. Also called construction documents.

Contract Specifications - A set of guidelines that a contractor must follow when conducting an asbestos abatement job.

Critical barrier means one or more layers of plastic sealed over all openings into a work area or any other similarly placed physical barrier sufficient to prevent airborne asbestos in a work area from migrating to an adjacent area.

Crocidolite - The strongest of the asbestos minerals. An asbestiform mineral of the amphibole group. It is of minor commercial value in the U.S. [Blue asbestos]

Damaged Friable Surfacing (Miscellaneous) Material - Friable surfacing (miscellaneous) ACM which has deteriorated or sustained physical injury such that the internal structure (cohesion) of the material is inadequate or, if applicable, which hasdelaminated such that the bond to the substrate (adhesion) is inadequate or which for any other reason lacks fiber cohesion or adhesion qualities. Such damage or deterioration may be illustrated by the separation of ACM into layers; separation of ACM from the substrate; flaking, blistering, or crumbling of ACM surface water damage; significant or repeated water stains, gouges, mars or other signs of physical injury on the ACM. Asbestos debris originating from the ACBM in question may also indicate damage (AHERA definition).

Damaged or Significantly Damaged Thermal System Insulation- Thermal system insulation on pipes, boilers, tanks, ducts, and other thermal system insulation equipment which has lost its structural integrity, or whose covering, in whole or in part, is crushed, water-stained, gouged, punctured, missing, or not intact such that it is not able to contain fibers. Damage may be further illustrated by occasional punctures, gouges, or other signs of physical injury to ACM, occasional water damage on the protective coverings/jackets; or exposed ACM ends or joints. Asbestos debris, originating from the ACBM in question may also indicate damage (AHERA definition).

Decontamination Unit - A series of connected rooms with polyethylene curtained doorways for the purpose of preventing contamination of areas adjacent to the work area. Usually constructed in a 5-chamber sequence of clean room, airlock, shower, airlock, dirty (or equipment) room.

Demand respirator - an atmosphere-supplying respirator that admits breathing air to the face piece only when a negative pressure is created inside the face piece by inhalation.

Demolition- means the wrecking or taking out of any load-supporting structural member and any related razing, removing, or stripping of asbestos products.

Dirty Room - The fifth or last chamber in the Decontamination Unit where workers will either collect equipment when entering a work area, or store equipment and remove protective clothing when exiting the work area. Also known as the Equipment Room.

Disturbance means activities that disrupt the matrix of ACM or PACM, crumble or pulverize ACM or PACM, or generate visible debris from ACM or PACM. In no event shall the amount of ACM or PACM so disturbed exceed that which can be contained in one glove bag or waste bag which shall not exceed 60 inches in length and width. Employee exposure means that exposure to airborne asbestos that would occur if the employee were not using respiratory protective equipment.

Dose-Response Effect - The relationship between the amount of pollutant a person is exposed to (dose) and the increased risk of disease (effect). Usually the greater the dose, the greater the effect.

Electrical Systems - The system of wires, lights, power generation equipment, and related facilities to produce, convey, and utilize electrical power in a building.

Electron Microscopy - A method of asbestos sample analysis which utilizes and electron beam to differentiate between fibers.

Encapsulation - The use of an agent to seal the surface (bridging encapsulant) or penetrate the bulk (penetrating encapsulant) of ACM.

Enclosure - A resilient structure, built (or sprayed) around ACM designed to prevent disturbance and contain released fibers.

EPA Regulations – Regulatory standards which cover emissions to the outside environment from a workplace and disposal of hazardous wastes from job sites. These include AHERA, ASHARA, and NESHAP.

Epidemiology - The study of causes, occurrence and distribution of diseases throughout a population.

Equipment Room-The part of the decontamination area workers remove and dispose of protective clothing and tools. This area or chamber is usually supplied with impermeable bags or container for disposal of contaminated protective clothing and equipment.

Errors and Omissions Insurance – A type of insurance, which protects professionals for mistakes they may make in contracted plans and recommendations.

Excursion Limit (EL)- A level of airborne fibers specified by OSHA as a short term excursion level. It is currently 1.0 fibers per centimeter (f/cc) of air, 30-minute time-weighted average, as measured by phase contrast rnicroscopy.

f/cc - Fibers per cubic centimeters of air. Units of measurement for asbestos in air.

Fiber- means a particulate form of asbestos, 5 micrometers or longer, with a length-to-diameter ratio of at least 3 to 1.

Fiber Containment – Enclosing or sealing off an area having airborne asbestos fibers present so that the fibers will not migrate resulting in contamination of other areas.

Fibrosis – A condition of the lungs caused by the inhalation of excessive amounts of fibrous dust marked by the presence of scar tissue.

Fireproofing - Spray, or trowel applied, fire resistant materials.

Friable Asbestos - Any materials that contain more than 1% asbestos and can be crumbled, pulverized, or reduced to powder by hand pressured when dry.

Full Face piece – A respirator which covers the wearer's entire face from the hairline to below the chin.
Functional Spaces - Spatially distinct units within a building, which contain identifiable populations of building occupants.
General Liability Insurance - A type of insurance which covers the insured for damage to property and person caused by his or her own negligence.
Globally Harmonized System of Classification & Labeling of Chemicals (GHS) – OSHA's new Hazard Communication standard incorporates portions of the UN Globally Harmonized System of Classification and Labeling of Chemicals (or GHS) and calls for **chemical manufacturers** to provide consistent labeling information that is intended to be easier to understand with pictograms. Major changes to the Hazard Communication Standard (HCS), found at 29 CFR 1910.1200 include:
--Revised criteria for classification of chemical hazards;
--Revised labeling provisions that include requirements for use of standardized signal words, pictograms, hazard statements, and precautionary statements; --A specified format for safety data sheets; and
--Related revisions to definitions and requirements for employee training on labels and safety data sheets
Glovebag- means not more than a 60 x 60 inch impervious plastic bag-like enclosure affixed around an asbestos-containing material, with glove-like appendages through which material and tools may be handled.
Glove-box (bag) – Plastic enclosure placed around a specific operation such as a valve to contain small areas of materials during asbestos removal.
Grade D Air - Breathing air which has been 19.5% - 23% oxygen, no more than 5 mg/m^3 of condensed hydrocarbons, no more than 10 ppm of carbon monoxide, no pronounced odor, and a maximum of 1,000 ppm carbon dioxide.
Ground Fault Interrupter – A device which automatically de-energizes any high voltage system component which has developed a fault in the ground line.
Half Mask – High Efficiency – A respirator which covers one-half of the wearer's face and is equipped with filters capable of screening out 99.97% of all particles larger than 0.3 microns.
Hazard Assessment -The interpretation and evaluation of physical assessment data in order to set abatement priorities and rank areas for response actions. These priorities and rankings are based on anticipated exposure to asbestos fibers.
Heat Cramps – Painful spasms of heavily used skeletal muscles such as hands, arms, legs, and abdomen which are sometimes accompanied by dilated pupils and weak pulse resulting from depletion of the salt content of the body.
Heat Exhaustion – A condition resulting from dehydration and/or salt depletion, or lack of blood circulation which is usually accompanied by fatigue, nausea, headache, giddiness, clammy skin, and a pale appearance.
Heat Stress – A bodily disorder associated with exposure to excessive heat.
Heat Stroke – The most severe of the heat stress disorders resulting from the loss of the body's ability to sweat which is characterized by hot dry skin,dizziness, nausea, severe headache, confusion, delirium, loss of

consciousness, confusion, and coma.

Heating, Ventilating, & Air Conditioning (HVAC) System - The system of pipes, ducts, and equipment (air conditioners, chillers, heaters,. boilers, pumps, fans) used to heat, cool, move, and filter air in a building. HVAC systems are also known as mechanical systems.

High Efficiency Particulate Air (HEPA) - A type of filter which is 99.97% efficient at filtering particles of 0.3 micrometers in diameter.

Homogeneous Sampling Area - An area of ACBM or suspect ACBM which appears similar throughout in terms of color, texture, and date of material application.

HVAC System – Heating, Ventilation, and Air Conditioning system usually found in large business and industrial facilities.

Indemnify - To pay for or pay back. Indemnification clauses in contracts are intended to cover the cost of judgments and/or legal defenses in the event of litigation.

Industrial Hygienist - A professional qualified by education, training, and experience to recognize, evaluate, and develop controls for occupational health hazards.

Latency Period – The time it takes a disease to manifest itself through physical symptoms or other forms of detection.

Liability - Being subject to legal action for one's behavior.

Local Education Agency (LEA)- Authority responsible for complying with AHERA. As defined in Section 198 of the Elementary and Secondary Education Act of 1965.

Logbook – An official record of all activities which occurred during an abatement project.

Lung Cancer - A malignant growth of abnormal cells in the lungs, specifically of the bronchi covering.

Macrophage - White blood cells, which attack foreign substances in the body. The release of enzymes from these cells as they attack indigestible particles, such as asbestos, contributes to the creation of scar tissue in the lung.

Make-up Air – Supplied or recirculated air to offset that which has already been exhausted from an area.

Management Plan - A plan for each LEA to control and manage ACBM (AHERA definition). Must be prepared by an EPA or state accredited Management Planner.

Management Planner - An individual that has completed an EPA or State approved course and passed an examination covering the development of management plans.

Manometer - A device that measures pressure difference; usually a U-shaped glass tube containing water or mercury.

Material safety data sheet (MSDS) A material safety data sheet (MSDS), safety data sheet (SDS), or product safety data sheet (PSDS) is an important component of product stewardship and workplace safety. It is intended to provide workers and emergency personnel with procedures for handling or working with that substance in a safe manner, and includes information such as physical data (melting point, boiling point, flash point, etc.), toxicity, health effects, first aid, reactivity, storage, disposal, protective equipment, and spill-handling procedures. MSDS formats can vary from source to source within a country depending on national requirements.

Maximum use concentration (MUC) means the maximum atmospheric concentration of a hazardous substance from which an employee can be expected to be protected when wearing a respirator, and is determined by the assigned protection factor of the respirator or class of respirators and the exposure limit of the hazardous substance.

Mechanical Systems - See HVAC systems.

Mesothelioma - A relatively rare form of cancer, which develops in the lining of the pleura or peritoneum with no known cure. It is almost always caused by exposure to asbestos.

Method 7400 - NIOSH sampling and analytical method for fibers using phase-contrast microscopy. Replaces method P7CAM 239.

Micrometer - One millionth of one meter

Mini-Enclosure- A polyethylene enclosure around a work area. Mini-enclosures are sealed enclosures used to protect the facility environment as a secondary means to help, or attempt to, contain fibers or debris generated during the work.

Miscellaneous Material - Interior building material on structural components, structural members or fixtures, such as floor and ceiling tiles, and does not include surfacing material or thermal system insulation (AHERA definition).

MSDS – Material Safety Data Sheet

Negative Pressure – An atmosphere created in a work area enclosure such that airborne fibers will tend to be drawn through the filtration system rather than leak out into the surrounding areas. The air pressure inside the work area is less than that outside the work area.

Negative Pressure Respirators - Respirators which function by the wearer breathing in air through a filter.

Negative Pressure Respirator Fit Check - A form of respirator fit evaluation in which the wearer covers the filters of a negative pressure, air-purifying respirator to check for leaks around the face seal.

NESHAP (National Emission Standards for Hazardous Air Pollutants): Section 112 of the Clean Air Act requires EPA to develop emission standards for hazardous air pollutants. In response, EPA published a list of hazardous air pollutants and promulgated the National Emission Standards for Hazardous Air Pollutants (NESHAP) Regulation 40 CFR subpart M, Part 61.

NIOSH - The National Institute for Occupational Safety and Health which was established by the Occupational Safety and Health Act of 1970. The official approving agencies for respiratory protective equipment who test and certify respirators.

Occurrence Insurance - A form of insurance in which a claim is allowed regardless of when the claim is filed. For asbestos insurance, the "occurrence" could be the time of first exposure.

NIOSH (National Institute for Occupational Safety and Health): The

National Institute for Occupational Safety and Health (NIOSH) is the federal agency responsible for conducting research and making recommendations for the prevention of work-related injury and illness. NIOSH is part of the Centers for Disease Control and Prevention in the Department of Health and Human Services.

NIOSH 7400: A light microscopy analytical method, also known as NIOSH Phase Contrast Microscopy [PCM] Method 7400.

NIOSH 9002: A polarized light microscopy (PLM) analytical method useful for the qualitative identification of asbestos and the semi-quantitative determination of asbestos content of bulk samples. The method measures percent asbestos as perceived by the analyst in comparison to standard area projections, photos, and drawings, or trained experience. The method is not applicable to samples containing large amounts of fine fibers below the resolution of the light microscope.

Operations and Maintenance Plan (O&M) - Specific procedures and practices developed for the interim control of asbestos-containing materials in buildings until it is removed.

OSHA - The Occupational Safety and Health Administration which was created by the Occupational Safety and Health Act of 1970; serves as the enforcement agency for safety and health in the workplace environment.

PACM means ``presumed asbestos containing material". In both the OSHA Construction Asbestos Standard (29 CFR 1926.1101) and the General Industry Asbestos Standard (29 CFR 1910.1001) PACM is defined as **thermal system insulation** (TSI) and surfacing material found in a building constructed no later than 1980. TSI is the material applied to pipes, fittings (joints, "Ts", elbows, valves, etc.), boilers, breechings, tanks, ducts or other structural components, generally to prevent heat loss or gain. **Surfacing material** refers to materials sprayed, troweled-on or otherwise applied to surfaces generally for acoustical, fireproofing, or other purposes. OSHA requires that building owners identify PACM in their buildings and treat the PACM as asbestos-containing materials (ACM) until the materials are proven not to contain asbestos.--Although not defined strictly as "PACM", both OSHA standards also require asphalt and vinyl flooring material installed no later than 1980 be "considered" and "treated" as asbestos-containing, until the building owner proves the flooring is not ACM. This includes not only the flooring material, but associated mastics and backings.

PAT Samples – Proficiency Analytical Testing of asbestos samples conducted through NIOSH for laboratories involved with the analysis of asbestos samples.

Permissible Exposure Limit (PEL) - A level of airborne fibers specified by OSHA as an occupational exposure standard for asbestos. It is currently 0.1 fibers per cubic centimeter of air, 8-hour time-weighted average, as measured by phase contrast microscopy.

PCM (Phase contrast microscopy): A light-enhancing microscope technology that employs an optical mechanism to translate small variations in phase into corresponding changes in amplitude, resulting in high-contrast images. Historically, this method was used to measure airborne fibers in occupational environments; however, it cannot differentiate asbestos fibers from other fibers.

PCMe (PCM-equivalent): This refers to chrysotile and amphibole structures identified through transmission electron microscopy (TEM) analysis that are equivalent to those that would be identified in the same sample through phase contrast microscopy analysis, with the main difference being that TEM additionally permits the specific identification of asbestos fibers. PCMe structures are asbestiform structures greater than 5 microns in length having at least a 3 to 1 length to width (aspect) ratio.

Personal air monitor: Also known as a low-flow or low-volume sample pump, this is an air sample pump that is portable so that it can be worn by a member of the sampling team during activity based sample.

Phase Contrast Microscopy (PCM) - An optical microscopic technique used for the counting of fibers in air samples, but which does not distinguish fiber types.

Physical Assessment - Assessing suspect material to determine the current condition of the material and the potential for future disturbance.

Plenum - A horizontal space designed to transport air in a building. Plenums are commonly the space between a dropped ceiling and the floor above.

Pleura-The thin membrane surrounding the lungs, and which lines the internal surface of the chest cavity.

Pleural Plaque - A fibrous thickening of the lining of the chest cavity. associated with asbestos exposure. **PLM (Polarized light microscopy):** A microscope technology that uses the polarity (or orientation) of light waves to provide better images than a standard optical microscope.

PPE: Personal protective equipment

Plumbing System - The system of pipes, valves, fittings and related components designed to convey liquid or gas fluids throughout a building. Some piping may also be part of the HVAC system.

Point Counting - A method of analyzing bulk samples whereby the sample is homogenized, placed on microscope slides and examined under a polarized light microscope. A point counting stage (or mechanical stage) and cross hair reticle are used for counting with only the particle(s) directly under the cross being counted (void space is not counted). A minimum of 400 counts should be made for each slide (several slides are examined).

Polarized Light Microscopy (PLM) - An optical microscopy technique for analyzing bulk samples for asbestos in which the sample is illuminated with polarized light (light which vibrates in only one plane) to distinguish between different types of asbestos fibers by their shape and unique optical properties.

Positive Pressure Respirators - Respirators which function by blowing air or providing pressured air to the wearer.

Positive Pressure Respirator Fit Check - A form of qualitative respirator fit testing in which the wearer covers the exhalation valve of a negative pressure, air-purifying respirator to check for leaks around the face seal.

Posting – Refers to caution or warning signs which should be posted in any area in which asbestos abatement is taking place, or where airborne fiber levels may present a health hazard.

Powered Air Purifying Respirator (PAPR) – Either a full face piece, helmet, or hooded respirator that has the breathing air powered to the wearer after it has been purified through a filter.
Pre-Construction Conference - A meeting held before any work begins between the contractor and the building owner at which time the job specifications are discussed and all details of the work agreed upon.
Pre-Employment Physical – Complete medical examination of an employee before the job begins to determine whether or not he/she is fit to perform the functions of their employment.
Pressure Demand Airline Devices – A respiratory protection device which has a regulator and valve design such that air enters the face piece only when the wearer breathes in.
Presumed ACM (PACM) – Defined in the OSHA Construction Industry Standards as thermal system insulation and sprayed on and/or troweled on, or otherwise applied surfacing material in buildings constructed no later than 1980. See PACM definition.
Project Designer - a person who has successfully completed the training requirements for an abatement project designer established by 40 U.S.C. 763.90(g).
Protection Factor (PF) - A number, which reflects the degree of protection provided by a respirator. It is calculated by dividing the concentration of contaminant outside the mask by the concentration inside the mask.
Protective Clothing – Protective, lightweight garments worn by workers performing asbestos abatement to keep gross contamination off the body.
Presumed ACM - Asbestos-containing thermal system insulation and surfacing materials found in a building constructed no later than 1980. (OSHA regulations)
Pulmonary - Pertaining to, or affecting the lungs, or some portion thereof.
Pulmonary Function Tests – A part of the medical examination required to determine the health status of a person's lungs.
Qualitative Fit Test - A method of testing a respirator's face-to-face piece seal by covering the inhalation or exhalation valves and either breathing in or out to determine the presence of any leaks.
Quality Assurance - A program for collecting and analyzing additional samples of suspect material to check on the reliability of procedures.
Quantitative Fit Testing - Testing the fit of a respirator by calculating concentrations of contaminants inside and outside the mask. This requires the use of instruments. **Rales** - Cracking sounds in the lower half of the lung; symptomatic of progressing asbestosis.
Random Sample - A sample drawn in such a way that there is no set pattern and is designed to give a true representation of the entire population or area.
Record Documents – Drawings and specifications, which should reflect the way a building was actually constructed (sometimes referred to as "as-built drawings")
Regulated area- means an area established by the employer to demarcate areas where airborne concentrations of asbestos exceed, or there is a reasonable possibility they may exceed, the permissible exposure limits.

Regulated Asbestos Containing Material (RACM) – As defined by the NESHAP regulations RACM includes: a) Friable asbestos material, b) Category I non-friable ACM that has become friable, c) Category I non-friable ACM that will be or has been subjected to sanding, grinding, cutting or abrading, or d) Category II non-friable ACM that has a high probability of becoming or has become crumbled, pulverized, or reduced

to powder by the forces expected to act on the material in the course of demolition or renovation operations regulated by subpart §61.141 of 40 CFR Part 61 (NESHAP Revision; Final Rule)

Removal-ACM is taken out or stripped from structures or substrates.

Renovation-Modifying an existing structure.

Repair-Any series of cutting, patching or replacing that improves damaged ACM to undamaged.

Respiration - The exchange of gases in the lungs.

Respiratory Protection Program - A set of procedures and equipment required by OSHA to be established by an employer, which provides for the safe use of respirators on their job sites.

Respiratory Tract - Respiratory Tract- The respiratory tract filters air throughout its length. From the entry to the nose (nose hairs), through the sinuses, the bronchi and bronchioles, down to the entry to the alveoli, different sized particles are filtered out all along the way. The larger particles are removed in the nose the sinuses. The smaller ones are filtered in the bronchioles, where they get stuck to the moist walls. The smallest particles, generally less than 10 microns in aerodynamic diameter, get past all the filtering mechanisms and reach the alveoli.

Self-contained breathing apparatus (SCBA*)* means an atmosphere-supplying respirator for which the breathing air source is designed to be carried by the user.

Serpentine: A name given to several members of a polymorphic group of magnesium silicate minerals—those having essentially the same chemistry but different structures or forms. Chrysotile asbestos is a member of the serpentine group.

Stationary air monitor: An air sample monitor that is placed in a single location and is not moved during one or more sampling events.

Shop Drawings - Detailed drawings of selected items used in the construction of a building that are drawn by the contract sible for designing the project or, but reviewed by the architect/engineer responsible for designing the project.

Significantly Damaged Friable Surfacing (Miscellaneous) Materials- Friable surfacing (miscellaneous) ACM in a functional space where damage is extensive and severe (AHERA definition).

Specifications - A written set of standards, procedures, and materials for the construction of a building.

Structural Member - Any load-supporting member such as beams and load supporting walls of a facility.

Submittals - Drawings or descriptive literature such as operating manuals transmitted to the building owner upon construction completion

Substrate - The material or existing surface located under or behind the asbestos-containing material.
Supplied Air Respirator – A respirator what has a central source of breathing air which is supplied to the wearer by way of an air line.
Surfactant – A chemical wetting agent added to water to improve its penetration abilities into asbestos-containing materials.
Surfacing Material - Material that is sprayed-on, troweled-on or otherwise applied to surfaces, such as acoustical plaster on ceilings and fireproofing materials on structural members, or other materials on surfaces for acoustical, fireproofing, or other purposes (AHERA definition).
Supplied-air respirator (SAR)- *or airline respirator* means an atmosphere- supplying respirator for which the source of breathing air is not designed to be carried by the user.
TEM (Transmission electron microscopy): A microscope technology and an analytical method to identify and count the number of asbestos fibers present in a sample. It uses the properties of electrons to provide more detailed images than polarized light microscopy (PLM). Capable of achieving a magnification of 20,000x.
Thermal System Insulation - Material applied to pipes, fittings, boilers, breeching, tanks, ducts, or other interior structural components to prevent heat loss or gain, or water condensation, or for other purposes.
Tort - A legal wrong, sometimes referred to as negligence.
Trachea - The main air tube into the lungs. Made up of cartilage and supported by cartilage rings, the trachea divides into two bronchi which lead into the lungs.
Transite™ - A trade name for asbestos cement wallboard and sheeting.
Transmission Electron Microscopy (TEM) - A method of microscopic analysis which utilizes an electron beam that is focused onto a thin sample. As the beam penetrates (transmits) through the sample, the difference in densities produces an image on a fluorescent screen from which samples can be identified and counted. Used for analyzing air samples for asbestos.
Tremolite - One of six naturally occurring asbestos minerals, Tremolite has few commercial uses.
TSCA (Toxic Substances Control Act): The Toxic Substances Control Act (TSCA) of 1976 was enacted by Congress to give EPA the ability to track the 75,000 industrial chemicals currently produced or imported into the United States.
TWA – Time-Weighted Average, as in air sampling.
Working drawings - A set of drawings, which reflect the intended construction and appearance of the building. Also known as building plans.
U.S. EPA - Environmental Protection Agency. Created in 1970, the U.S. EPA is the federal promulgator and enforcement agency for environmental regulations.
User Seal Check - Procedure for ensuring the seal of a negative pressure, air-purifying respirator by covering the filters and inhaling, then the exhalation valve and exhaling to check for leaks around the face seal.